Deepak Pandey
Fiber-Based Optical Resonators

Also of Interest

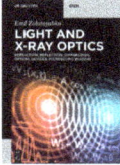

Light and X-Ray Optics
Refraction, Reflection, Diffraction, Optical Devices, Microscopic Imaging
Zolotoyabko, 2023
ISBN 978-3-11-113969-2, e-ISBN (PDF) 978-3-11-114010-0,
e-ISBN (EPUB) 978-3-11-114089-6

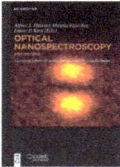

Optical Nanospectroscopy
Applications
Meixner, Fleischer, Kern, Sheremet, McMillan (Eds.), 2022
ISBN 978-3-11-044289-2, e-ISBN (PDF) 978-3-11-044290-8,
e-ISBN (EPUB) 978-3-11-043498-9

Electrons in Solids
Mesoscopics, Photonics, Quantum Computing, Correlations, Topology
Bluhm, Brückel, Morgenstern, Plessen, Stampfer, 2019
ISBN 978-3-11-043831-4, e-ISBN (PDF) 978-3-11-043832-1,
e-ISBN (EPUB) 978-3-11-042929-9

Optical Electronics
An Introduction
Yan, in Cooperation with Tsinghua University Press, 2019
ISBN 978-3-11-050049-3, e-ISBN (PDF) 978-3-11-050060-8,
e-ISBN (EPUB) 978-3-11-049800-4

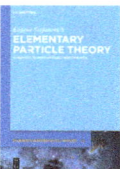

Elementary Particle Theory
Volume 2: Quantum Electrodynamics
Stefanovich, 2018
ISBN 978-3-11-049089-3, e-ISBN (PDF) 978-3-11-049320-7,
e-ISBN (EPUB) 978-3-11-049143-2

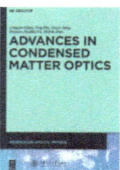

Advances in Condensed Matter Optics
Chen, Dai, Jiang, Jin, Liu, 2014
ISBN 978-3-11-030693-4, e-ISBN (PDF) 978-3-11-030702-3,
e-ISBN (EPUB) 978-3-11-038818-3

Deepak Pandey

Fiber-Based Optical Resonators

Cavity QED, Resonator Design and Technological Applications

DE GRUYTER

Author
Dr. Deepak Pandey
Menlo Systems GmbH
Bunsenstraße 5
82152 Martinsried
Germany
d.pandey@menlosystems.com

ISBN 978-3-11-063623-9
e-ISBN (PDF) 978-3-11-063626-0
e-ISBN (EPUB) 978-3-11-063630-7

Library of Congress Control Number: 2023946887

Bibliographic information published by the Deutsche Nationalbibliothek
The Deutsche Nationalbibliothek lists this publication in the Deutsche Nationalbibliografie;
detailed bibliographic data are available on the Internet at http://dnb.dnb.de.

© 2024 Walter de Gruyter GmbH, Berlin/Boston
Cover image: alwyncooper / E+ / Getty Images
Typesetting: VTeX UAB, Lithuania
Printing and binding: CPI books GmbH, Leck

www.degruyter.com

To my friends and family

Preface

Fiber-optic technology has revolutionized the field of optics, imaging, and telecommunication by making light transportation with low losses and precise mode control possible. The availability of low-loss fiber has paved the way for high data transmission and high bandwidth capacity. It plays a pivotal role in modern communication and network systems. Moreover, optical fibers have led to the emergence of various new areas, including non-linear fiber optics, fiber sensor technology, fiber-based lasers, and optical amplifiers.

With the advancement in fiber-optic technology, there is a strong research trend toward developing fiber-based devices, including fiber-based optical resonators. These resonators originate in the quest to develop miniature microresonators to integrate them into fiber optical networks. In recent years, research efforts have intensified due to the widespread applications of optical resonators in fields such as laser physics, atomic physics, and particularly in the quantum technology domain. Miniaturized fiber resonators are particularly attractive for quantum technology because they enable the highly confined manipulation of light and matter properties at the single-particle level. Cavity quantum electrodynamics (Cavity-QED) is the subject exploring this domain of scientific curiosity.

I have written this brief account of fiber-based resonators to help the reader become acquainted with optical resonators and their miniaturized version in a fiber-based framework. It is deliberate here to describe only fiber-based systems with fiber acting only as a passive channel for light transmission. I have taken the liberty only to take specialized fiber systems to cover topics of interest and skipped the cases in which fiber-based resonator systems are directly used as a gain or non-linear medium. The choice of topics here is manifestly governed by the research topics, I have encountered and been exposed to while working with cavity electrodynamical systems. I worked directly on the development and research of fiber-based Fabry-Perot micro-resonators and, therefore, explained these systems extensively.

In the current capacity of the book, I am unable to cover the broad field of all miniaturized micro and nano-resonator systems, which are currently being researched and developed in different laboratories.

The pages of this book are enriched with the wisdom of scholars, experts, and visionaries who have dedicated their lives to advancing the field of cavity-QED, optical resonators, and sensors. Their contributions have enriched the content of this book, ensuring that we can present a comprehensive experience of fiber-based resonator systems.

Chapter 1 introduces the basics of electromagnetic wave propagation in a bounded medium. Chapter 2 systematically introduces the basics of optical resonators and some of their applications in precision measurements. Chapter 3 details the fiber microresonators and focuses on their design and fabrication techniques. Chapter 4 introduces the cavity quantum electrodynamics and application of fiber microresonators in the

https://doi.org/10.1515/9783110636260-201

field of cavity-QED. Chapter 5 briefly discusses the future perspective of fiber-resonator applications in quantum technology.

I want to thank all the research groups whose work I have freely used to describe fiber-based microresonator systems. I thank all the friends and colleagues involved during the writing process. Many people have proofread sections of the book at different levels and have provided helpful remarks and suggestions. Especially, thanks to my friend Murtaza Ali Khan for being deeply involved during the final stages of the writing. Thanks also go to people whose research work has contributed directly or indirectly and those who have contributed to generating some figures for the book, as my colleagues Carlos Saavedra, Maximillian Ammenwerth, Jose Gallego, Michael Kubista, David Röser, Madhavakkannan Saravanan, Santosh Surendra and other members with whom I worked at the University of Bonn on the topic of fiber-based Fabry-resonators. Also, other friends, Srihari Srinivasan, Deepika Bollimpalli, and Hasan Kazi, for helping at various stages of the book. Big thanks to my wife, Chandreyee Maitra, for keeping me on the disciplined track of writing and also for her help in proofreading the chapters.

I am deeply grateful for the opportunity to share this book with you. I sincerely hope it will be concise yet detailed enough to jump-start in the field of miniature fiber-based resonators.

Yours sincerely, Deepak Pandey

Contents

1 Basics of electromagnetic waves

In the 19th century, one of the most important breakthroughs in physics occurred when the separate theories of electrostatics and magnetism were unified. This unification also established a direct link between the laws of electromagnetism and the principles governing the propagation of light. James Clerk Maxwell, a renowned Scottish physicist, played a pivotal role in formulating the unified electromagnetic wave theory. This theory is encapsulated by a set of four differential equations known as Maxwell's equations, which underpin the summary of almost all electromagnetic (EM) phenomena and EM-wave propagation, including light. These equations serve as the cornerstone for explaining a wide range of electromagnetic phenomena in electronics, optics, and communication sciences. This chapter provides a brief overview of Maxwell's equations and their practical applications in comprehending the propagation of electromagnetic waves under specific boundary conditions. The chapter includes examples of electromagnetic wave propagation in waveguide systems and fiber optic waveguides, illustrating the practical implications of Maxwell's equations in these contexts.

1.1 Electromagnetism

The scientific theory of unified electromagnetism evolved from years of work by mathematicians and physicists in formulating individual theories of electrostatics and magnetic phenomena [4]. To name a few which directly influenced the work of James Clerk Maxwell were the pioneering works of Carl Friedrich Gauss (1775–1855), Andere Ampere (1775–1836), and Michael Faraday (1791–1867). They established the fact that:
- The electric flux through an enclosed surface is related to the volumetric charge density, ρ (Figure 1.1).

$$\boxed{\nabla \cdot \mathbf{E} = \frac{\rho}{\epsilon_0}} \tag{1.1}$$

This law is named after the German mathematician and physicist, Carl Friedrich Gauss, and holds true for any closed surface, regardless of its shape or size. The integral form of Gauss law can be expressed in terms of total flux through a surface $\int_s \mathbf{E} \cdot d\mathbf{A} = \int_V \frac{\rho \, dV}{\epsilon_0}$, and using relation $\int_s \mathbf{E} \cdot d\mathbf{A} = \int_V (\nabla \cdot \mathbf{E}) \, dV$, one obtains equation (1.1).
- A changing magnetic field produces an electric field and, therefore, induces electricity in a metallic coil through which the magnetic flux transverse (Figure 1.2). This law known as Faraday's law was discovered by the English scientist Michael Faraday in the early 19th century and describes the relationship between a chang-

https://doi.org/10.1515/9783110636260-001

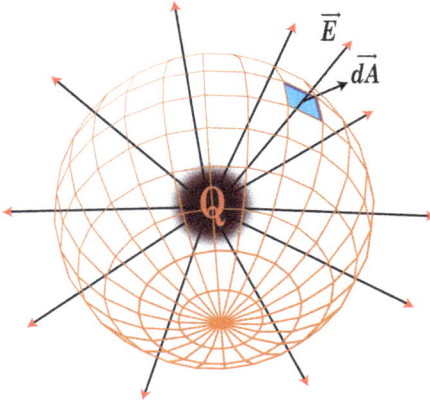

Figure 1.1: Depiction of Guass law for electric field lines due to an enclosed charge inside a sphere. The electric flux directly depends the total charge enclosed: $\int_s \mathbf{E} \cdot d\mathbf{A} = \frac{Q}{\epsilon_0}$.

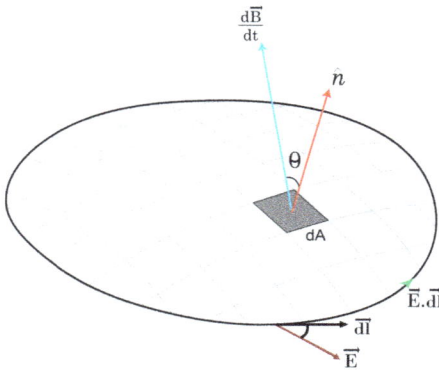

Figure 1.2: The induced electromotive force (EMF) on a closed loop is equal to the line integral of the electric field. Faraday's law states that the induced EMF is related to the rate of change of total magnetic flux, $\oint \mathbf{E} \cdot d\mathbf{l} = -\frac{d}{dt}(\int_s \nabla \times \mathbf{B} \cdot d\mathbf{A})$.

ing magnetic field and the induction of an electromotive force (EMF) or voltage in a conducting loop or coil.

$$\boxed{\nabla \times \mathbf{E} = -\frac{\partial \mathbf{B}}{\partial t}}$$ (1.2)

The integral form of Faraday law relates induced EMF(e) in a closed loop due to the changing magnetic flux as $e = -\frac{d\phi}{dt} = -\frac{d(\int_s \mathbf{B} \cdot d\mathbf{A})}{dt}$. Combining this with relations $e = \oint \mathbf{E} \cdot d\mathbf{l}$ and $\oint \mathbf{E} \cdot d\mathbf{l} = \int_s \nabla \times \mathbf{E} \cdot d\mathbf{A}$ produces equation (1.2).

– The magnetic field in a closed loop circuit depends on the current through the loop. This law known as Ampere's law and quantifies the relationship between the distribution of electric currents and the resulting magnetic field. In its differential form, the curl of the magnetic field depends on the associated current density J (Figure 1.3).

$$\boxed{\nabla \times \mathbf{B} = \mu_0 \mathbf{J}}$$ (1.3)

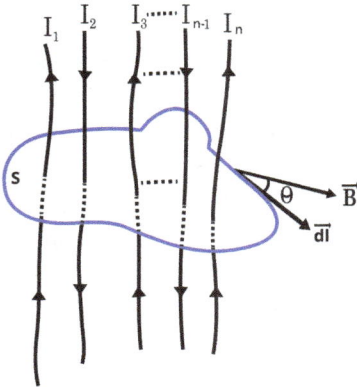

Figure 1.3: The line integral of the magnetic field for a closed loop is related to the enclosed current sources, $\oint \mathbf{B} \cdot d\mathbf{l} = \mu_0 \int_s \mathbf{J} \cdot d\mathbf{A}$.

The integral form of Ampere's law relates the total line integral of magnetic field in a closed loop to the total current threading through the enclosed surface, $\oint \mathbf{B} \cdot d\mathbf{l} = \mu_0 \sum_i I_i = \mu_0 \int_s \mathbf{J} \cdot d\mathbf{A}$. Combining this with Stoke's theorem $\oint \mathbf{B} \cdot d\mathbf{l} = \int_s \mathbf{B} \cdot d\mathbf{A}$ yields equation (1.3).

Here, ϵ_0 and μ_0 are the electric permittivity and magnetic permeability in a vacuum. $\nabla \cdot ()$ and $\nabla \times ()$ are the divergence and the curl operators.

These discoveries made it clear that the two seemingly different phenomena, electricity and magnetism, are both produced by electric charges depending on whether they are stationary or in motion. With the realization of electromagnetism as a unified theory, an explicit mathematical description of electromagnetism emerged through the meticulous theoretical work of Maxwell. Simple differential equations derived from his work contained all the essence of the electromagnetic phenomenon. Furthermore, Maxwell expanded upon the existing theory by introducing an additional term, known as the displacement current to Ampere's law. This addition proved crucial for a comprehensive understanding of electromagnetic behavior. After further research and simplification by British physicist, Oliver Heaviside, these equations were finally condensed into the currently renowned four Maxwell electromagnetic wave equations, which present a concise and elegant description of electromagnetic waves, including light and the associated phenomena in wave optics.

1.2 Electromagnetic waves

An electromagnetic wave (EM-wave) consists of oscillating electric and magnetic fields at each point in space and time. These fields exhibit sinusoidal patterns, much like the oscillation of particles in a mechanical wave traveling through a medium. EM-wave is created by the interaction of electric charges and magnetic fields following the principles of electromagnetism. EM-waves are characterized by their frequency (ν) and wavelength (λ), which are related by the equation: $c = \nu\lambda$, where c is the speed of light in a vacuum. This equation highlights that electromagnetic waves travel at the speed of light.

The electromagnetic spectrum encompasses a wide range of frequencies and wavelengths, including radio waves, microwaves, infrared waves, visible light, ultraviolet waves, X-rays, and Gamma rays. Each segment of the spectrum corresponds to a different range of frequencies and wavelengths, with visible light being the portion that can be detected by the human eye.

As shown in Figure 1.4, the electric field vector **E** and magnetic field vector **B** are orthogonal to the wave propagation direction and their amplitudes sinusoidally vary in space and time. This description of EM-wave directly arises from the four fundamental laws of electromagnetism. They describe a deeper connection between the following four basic quantities of electromagnetism [3]:

– Displacement vector inside a medium $\mathbf{D} = \epsilon_0\mathbf{E} + \mathbf{P}$
– Magnetization vector $\mathbf{B} = \mu_0(\mathbf{H} + \mathbf{M})$
– Charge density ρ
– Current density $\mathbf{J} = \sigma\mathbf{E}$

where **P** and **M** are defined as the polarizability and the magnetization and are related to electric and magnetic fields via electric (χ_e) and magnetic susceptibility (χ_m), respectively.

$$\mathbf{P} = \epsilon_0\chi_e\mathbf{E}; \quad \mathbf{M} = \chi_m\mathbf{H} \tag{1.4}$$

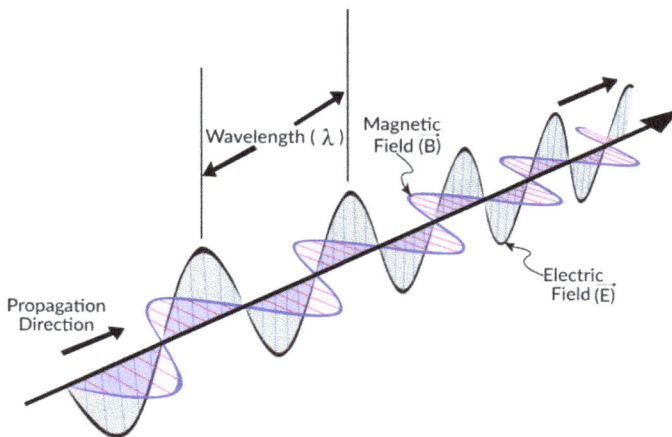

Figure 1.4: Electromagnetic wave propagation in a vacuum can be considered as oscillating electric and magnetic fields with a given amplitude in two orthogonal axes and perpendicular to the propagation axis.

Note that with equation (1.4), Displacement and magnetization vectors can also be written in terms of vacuum and relative permittivity and permeability, respectively,

$$\mathbf{D} = \epsilon_0 \left(1 + \chi_e\right) \mathbf{E} = \epsilon_0 \, \epsilon_r \, \mathbf{E} \tag{1.5}$$

$$\mathbf{B} = \mu_0 \left(1 + \chi_m\right) \mathbf{H} = \mu_0 \, \mu_r \, \mathbf{E} \tag{1.6}$$

where $\epsilon_r = 1 + \chi_e$ is the relative permittivity and $\mu_r = 1 + \chi_m$ is the relative permeability.

Maxwell's equation establishes relations between four quantities $\mathbf{D}$, $\mathbf{B}$, ρ and $\mathbf{J}$. We will see in the next section that Maxwell's equations can be used to derive a set of differential equations whose solution produces classical wave equations with electric and magnetic field vectors recognized as oscillating variables and speed of light as a fundamental constant defined in terms of the permittivity of the medium, producing a velocity constant $\frac{1}{\sqrt{\mu_0 \epsilon_0}}$ in a vacuum, equal to the speed of light.

1.3 Maxwell's equations

Maxwell's equations are a set of four fundamental equations that describe the behavior of electric and magnetic fields and their interplay with electric charges and currents. These equations concisely describe the relationship between electromagnetic quantities, providing the complete description of electromagnetic phenomena under various conditions. The four Maxwell's equations in the differential form relating the fundamental quantities of electromagnetism are:

$$\nabla \cdot \mathbf{D} = \rho \tag{1.7}$$

$$\nabla \times \mathbf{E} = -\frac{\partial \mathbf{B}}{\partial t} \tag{1.8}$$

$$\nabla \cdot \mathbf{B} = 0 \tag{1.9}$$

$$\nabla \times \mathbf{H} = \mathbf{J} + \frac{\partial \mathbf{D}}{\partial t} \tag{1.10}$$

The most intriguing feature of Maxwell's equations is revealed by obtaining a wave equation with $\mathbf{E}$ and $\mathbf{B}$ being the oscillating terms of the propagating wave as follows.

By applying $\nabla \times$ operator to equation (1.8),

$$\nabla \times (\nabla \times \mathbf{E}) = -\mu_0 \frac{\partial (\nabla \times \mathbf{H})}{\partial t}. \tag{1.11}$$

Using the following vector identity,

$$\nabla \times (\nabla \times \mathbf{A}) = \nabla(\nabla \cdot \mathbf{A}) - \nabla^2 \mathbf{A}, \tag{1.12}$$

which in combination with equation (1.9), produces the following wave equation. Note $\nabla \cdot \mathbf{D} = 0$:

$$\nabla^2 \mathbf{E} - \frac{1}{c^2} \frac{\partial^2 \mathbf{E}}{\partial t^2} = 0, \tag{1.13}$$

and similarly for the magnetic field

$$\nabla^2 \mathbf{B} - \frac{1}{c^2} \frac{\partial^2 \mathbf{B}}{\partial t^2} = 0, \tag{1.14}$$

where c is the speed of light in a given medium of refractive index n. The wave equation for both electric and magnetic fields in a vacuum reads as follows:

$$\nabla^2 \mathbf{E} - \frac{1}{c_0^2} \frac{\partial^2 \mathbf{E}}{\partial t^2} = 0; \quad \nabla^2 \mathbf{B} - \frac{1}{c_0^2} \frac{\partial^2 \mathbf{B}}{\partial t^2} = 0. \tag{1.15}$$

The speed of light in vacuum $c_0 = \frac{c}{n} = \frac{1}{\sqrt{\epsilon_0 \mu_0}}$.

1.3.1 Energy transfer and Poynting vector

The energy transported by EM-wave per unit time per unit area is an important characteristic parameter of the EM field. This is represented by the Poynting vector as follows [3]:

$$\mathbf{S} = \frac{1}{\mu_0} (\mathbf{E} \times \mathbf{B}). \tag{1.16}$$

Thus, the Poynting vector is a mathematical quantity used in electromagnetism to describe the direction and rate of energy flow in an electromagnetic field. It represents the power per unit area, also known as the energy flux, carried by an electromagnetic wave. The Poynting vector has both magnitude and direction. The magnitude of the Poynting vector represents the power per unit area carried by the electromagnetic wave. Its direction gives the direction of energy flow, which is perpendicular to both the electric and magnetic field vectors (Figure 1.5). The energy flow is in the direction of the Poynting vector.

The expression equation (1.16) can be understood by calculating the work done by the EM-wave on some charge configuration inside volume V, which is as follows:

$$\frac{\partial W}{\partial t} = \int_V (\mathbf{E} \cdot \mathbf{J}) d\tau \tag{1.17}$$

Using Maxwell's equations, the above expression takes the following form:

$$\frac{\partial W}{\partial t} = -\frac{d}{dt} \left(\int_V \left(\frac{1}{2} \epsilon_0 \mathbf{E}^2 + \frac{1}{\mu_0} \mathbf{B}^2 \right) d\tau \right) - \frac{1}{\mu_0} \int_S (\mathbf{E} \times \mathbf{B}) \cdot d\mathbf{s} \tag{1.18}$$

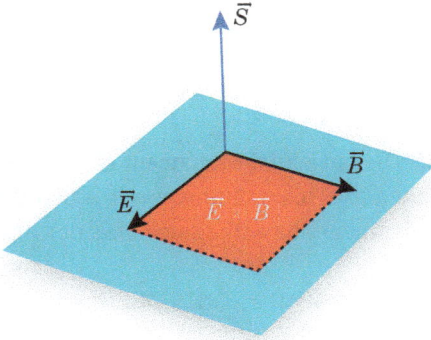

Figure 1.5: Poynting vector **S** represents the energy transported per unit time per unit area in a plane orthogonal to the plane containing **E** and **B** vectors.

Here, $\int_V$ and $\int_S$ denotes the volume and surface integral over a selected region, within which the charges are enclosed.

Therefore, the work done on a charge configuration by EM-field interaction is equal to the reduction in the rate of stored energy and the energy flowing out of a given surface. For this reason, the last expression in equation (1.18) represents the energy transported by the EM-fields per unit area, per unit time, and is referred to as the Poynting theorem. The Poynting theorem is an energy conservation theorem for a time-dependent electromagnetic field. The Poynting theorem is an energy conservation theorem for a time-dependent electromagnetic field.

1.3.2 Maxwell's equations under boundary condition

One of the fascinating applications of Maxwell's equations is their use in theoretical analyses of electromagnetic (EM) wave propagation within confined/bounded media. Examples of such propagation include EM waves traveling through hollow metallic waveguides, coaxial cables at microwave frequencies, and optical fibers for light transmission. Guided propagation of EM waves forms the foundation of optical communication systems.

To accurately describe the behavior of electromagnetic fields at boundaries or interfaces between different media, Maxwell's equations must be applied alongside specific boundary conditions. These conditions establish constraints on the electric field (**E**) and magnetic field (**B**) at these interfaces, ensuring continuity and accounting for the properties of the materials involved. The boundary conditions for the electric and magnetic fields are considered separately. For example, in some cases:

- *Boundary conditions for the electric field:* The tangential component of the electric field must be continuous across a boundary. This means that the electric field vectors on both sides of the boundary must have the same direction and magnitude along the boundary.

The normal component of the electric displacement field $\mathbf{D} = \epsilon_0\mathbf{E} + \mathbf{P}$, where ϵ_0 is the electric permittivity constant and $\mathbf{P}$ is the polarization must be continuous across a boundary. This condition accounts for the effects of electric charges and polarization of materials.

– *Boundary conditions for the magnetic field:* The tangential component of the magnetic field must be continuous across a boundary. This condition ensures that the magnetic field vectors on both sides of the boundary have the same direction and magnitude along the boundary.

The normal component of the magnetic field intensity $\mathbf{B} = \mu_0(\mathbf{H} + \mathbf{M})$, where μ_0 is the magnetic constant and $\mathbf{H}$ is the magnetic field, and $\mathbf{M}$ is the magnetization must be continuous across a boundary. This condition considers the effects of magnetic currents and magnetization of materials.

In addition to these boundary conditions, specific conditions can arise in certain scenarios:

Perfect conductors: At the boundary of a perfect conductor, the electric field must be zero, meaning that the tangential component of the electric field is perpendicular to the surface of the conductor.

Dielectric interfaces: At the boundary between two dielectric materials with different electric constants (ϵ), the tangential components of the electric field are related by the ratio of the electric constants: $E_{1t}/\epsilon_1 = E_{2t}/\epsilon_2$, where E_{1t} and E_{2t} are the tangential components of the electric field on each side of the boundary.

These boundary conditions play a crucial role in maintaining the integrity and coherence of electromagnetic fields across interfaces, allowing for the smooth transition of waves between different media. They provide essential constraints and help ensure the consistency and continuity of electromagnetic fields in various situations and material interfaces. The boundary conditions, combined with Maxwell's equations, allow for the determination of the behavior of electric and magnetic fields at interfaces and boundaries, enabling the analysis of electromagnetic wave propagation, reflection, and transmission in different media.

1.3.3 Waveguide and waveguide modes

Guiding electromagnetic waves with appropriate boundary conditions in a confined geometry is the basis of current-day electromagnetic and fiber optic communication technologies. Maxwell's equation underpins the methods to calculate reflection and refraction within the guiding medium. In the following section, an example of a direct application of Maxwell's equations for waveguides is discussed. The specific case of light propagation in the fiber optic and confinement of light in an optical resonator is discussed in further sections.

A waveguide is a structure or device used to guide and confine electromagnetic waves along a specific path. It is designed to propagate and control the transmission of electromagnetic energy, typically in the form of microwaves or radio waves. Waveguides are commonly used in various fields, including telecommunications, radar systems, microwave engineering, and high-frequency applications. Waveguides are typically constructed as hollow metallic or dielectric tubes, although other shapes and materials can be used depending on the specific requirements. The inner walls of the waveguide are highly reflective to the electromagnetic waves, allowing them to bounce back and forth within the waveguide without significant energy loss. This reflection is often achieved through the principle of total internal reflection, where the waveguide walls act as mirrors for the waves. The main purpose of a waveguide is to provide a controlled pathway for electromagnetic waves, allowing them to be efficiently transmitted, directed, and manipulated. By confining the waves within the waveguide, they can be guided over long distances while minimizing losses due to radiation and external interference. Waveguides support the propagation of different modes, which represent distinct patterns of electric and magnetic fields within the structure. These modes depend on the geometry, dimensions, and boundary conditions of the waveguide. The selection of a specific mode depends on the desired performance and functionality of the waveguide system.

Waveguide modes refer to the different possible patterns of electromagnetic wave propagation within a waveguide. The modes of a waveguide are distinct patterns of electric and magnetic fields that can exist within the waveguide. These modes are characterized by their unique field distributions and corresponding propagation characteristics. The number and types of modes depend on the geometry and dimensions of the waveguide. Here are some key concepts related to waveguide modes.

Transverse Electric (TE) modes: TE-modes are electromagnetic wave modes in which the electric field is purely transverse (perpendicular) to the direction of propagation. In these modes, the magnetic field has both transverse and longitudinal components. The subscript "m,n" is used to represent the mode indices, where "m" denotes the number of half-wavelength variations of the electric field across the width of the waveguide, and "n" represents the number of half-wavelength variations along the length.

Transverse Magnetic (TM) modes: TM-modes are electromagnetic wave modes in which the magnetic field is purely transverse to the direction of propagation. In these modes, the electric field has both transverse and longitudinal components. Similar to TE modes, the mode indices "m,n" are used to represent the variations of the magnetic field across the width and length of the waveguide.

Hybrid modes: Hybrid modes, also known as TE-TM modes or EH-modes, have both electric and magnetic field components in both transverse and longitudinal directions. These modes possess characteristics of both TE and TM modes and are generally more complex in nature.

The propagation characteristics of waveguide modes, such as the cutoff frequency, dispersion, and attenuation, depend on the dimensions and boundary conditions of the waveguide. The mode with the lowest cutoff frequency is called the fundamental mode, while higher-order modes have higher-cutoff frequencies and exhibit more complex field patterns.

The selection of a specific waveguide mode depends on the desired performance and functionality of the waveguide system. One can analyze the field distributions, dispersion properties, and other factors to design and optimize waveguides for specific applications.

Understanding waveguide modes is crucial for the efficient transmission and control of electromagnetic waves within waveguide structures, allowing for the manipulation and utilization of electromagnetic energy in various technological applications. For an EM-wave propagation through a metallic waveguide, assuming it is a perfect conductor, the typical boundary conditions are:

1) The tangential component of the electric field should be equal to zero.
2) The normal derivative of the tangential component of the magnetic field should be equal to zero.

For a typical hollow waveguide, the wave propagation involves guiding the EM fields inside a hollow metal pipe. In an ideal case, for a monochromatic wave propagating inside the waveguide and considering the above boundary conditions, the following general electric and magnetic field vectors can be written for the field inside the waveguide

$$\mathbf{E}(\mathbf{x}, \mathbf{y}, \mathbf{z}, \mathbf{t}) = \mathbf{E_0}(\mathbf{x}, \mathbf{y}, \mathbf{z}) \exp i(kz - \omega t) \qquad (1.19)$$

$$\mathbf{B}(\mathbf{x}, \mathbf{y}, \mathbf{z}, \mathbf{t}) = \mathbf{B_0}(\mathbf{x}, \mathbf{y}, \mathbf{z}) \exp i(kz - \omega t) \qquad (1.20)$$

where $\mathbf{E_0}(\mathbf{x}, \mathbf{y}, \mathbf{z})$ and $\mathbf{B_0}(\mathbf{x}, \mathbf{y}, \mathbf{z})$ are three-dimensional vectors with components (E_x, E_y, E_z) and (B_x, B_y, B_z).

As there are no free charges present inside the waveguide, Maxwell's equations are as follows:

$$\nabla \cdot \mathbf{E} = 0 \qquad (1.21)$$

$$\nabla \times \mathbf{E} = -\frac{\partial \mathbf{B}}{\partial t} \qquad (1.22)$$

$$\nabla \cdot \mathbf{B} = 0 \qquad (1.23)$$

$$\nabla \times \mathbf{B} = \frac{1}{c^2} \frac{\partial \mathbf{E}}{\partial t} \qquad (1.24)$$

Now one can use the two of the Maxwell's equations (equation (1.22) and equation (1.24)) and do further simplification to obtain x- and y-field components in terms of E_z and B_z:

$$\mathbf{E}_x = \frac{i}{k_w{}^2}\left(k\frac{\partial E_z}{\partial x} + \omega\frac{\partial B_z}{\partial y}\right); \quad \mathbf{E}_y = \frac{i}{k_w{}^2}\left(k\frac{\partial E_z}{\partial y} - \omega\frac{\partial B_z}{\partial x}\right), \tag{1.25}$$

$$\mathbf{B}_x = \frac{i}{k_w{}^2}\left(k\frac{\partial B_z}{\partial x} - \frac{\omega}{c^2}\frac{\partial E_z}{\partial y}\right); \quad \mathbf{B}_y = \frac{i}{k_w{}^2}\left(k\frac{\partial B_z}{\partial y} + \frac{\omega}{c^2}\frac{\partial E_z}{\partial x}\right), \tag{1.26}$$

where $k_w = (\omega/c)^2 - k^2$.

Using these relations with the other two Maxwell's equations (equations (1.21) and equation (1.23)), one obtains two uncoupled differential equations for z-components:

$$\left[\frac{\partial^2}{\partial x^2} + \frac{\partial^2}{\partial y^2} + \left(\frac{\omega}{c}\right)^2 - k^2\right]E_z = 0 \tag{1.27}$$

$$\left[\frac{\partial^2}{\partial x^2} + \frac{\partial^2}{\partial y^2} + \left(\frac{\omega}{c}\right)^2 - k^2\right]B_z = 0 \tag{1.28}$$

Therefore, solving the above differential equations for z-components with the boundary conditions allows obtaining the complete description of the field components for waveguide propagation.

The two specific cases of waveguide propagation are for:

- $E_z = 0$: This is called as Transverse electric (TE-wave): $\mathbf{E} = (E_x, E_y, 0)$ and $\mathbf{B} = (B_x, B_y, B_z)$
- $B_z = 0$: This is called as Transverse magnetic (TM-wave): $E = (E_x, E_y, E_z)$ and $B = (B_x, B_y, 0)$

It can be established from Maxwell's equations that both $E_z = 0$ and $B_z = 0$ (which is known as TEM$_{00}$) mode is not possible inside a hollow metallic waveguide. However, a coaxial cable with one inner and outer conductor configuration and the in-between space filled with a dielectric medium allows for the propagation of the lowest fundamental mode, i. e., TEM$_{00}$ mode.

1.3.4 Rectangular waveguide

Typical waveguides used for guided EM-wave transmission are usually rectangular (Figure 1.6) or mostly cylindrical in shape. The design considerations of these waveguides are based on solving Maxwell's differential equations with boundary conditions. Assuming a hollow rectangular metal waveguide with dimensions $a \times b$ as shown in Figure 1.7. There are two distinct sets of possible solutions, transverse-magnetic (TM) with $B_z = 0$ and transverse-electric (TE) with $E_z = 0$. Considering TE-wave propagation, one needs to solve the wave equation for B_z (equation (1.28)). The remaining field components are then calculated from equations (1.25)–(1.26).

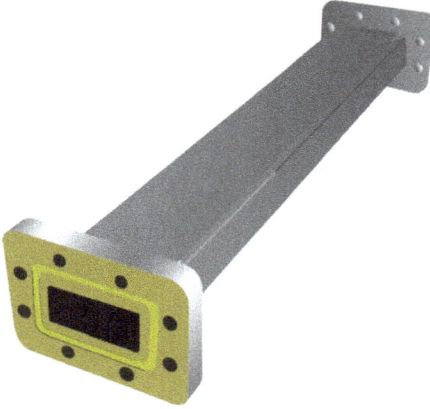

Figure 1.6: A rectangular metal waveguide.

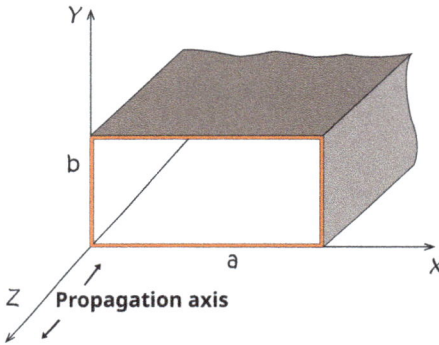

Figure 1.7: Hollow metallic waveguides are used for guiding EM-wave with microwave frequencies. The figure shows a rectangular waveguide with dimensions a and b along the two orthogonal x- and y-axes and extension along the propagation direction (z-axis). Maxwell's differential equations can be solved to obtain the limits on the frequencies that can propagate with minimum losses.

We use separation of variables method for solving the differential equation (1.28). Assuming B_z as the following product function:

$$B_z(x,y) = X(x)Y(y), \tag{1.29}$$

the differential equation have the following form:

$$\frac{1}{X}\left(\frac{d^2X}{dx^2}\right) + \frac{1}{Y}\left(\frac{d^2Y}{dy^2}\right) + \left(\left(\frac{\omega}{c}\right)^2 - k^2\right)B_z = 0. \tag{1.30}$$

It is clear that the x- and y-dependent terms must be constant for the above equation. This condition enforces two unknown constants with the following relations:

$$\frac{1}{X}\left(\frac{d^2X}{dx^2}\right) = -k_x^2, \quad \frac{1}{X}\left(\frac{d^2X}{dx^2}\right) = -k_y^2, \tag{1.31}$$

and the following relation between the unknown constants (k_x and k_y) and the EM-wave constants:

$$-k_x^2 - k_y^2 + \left(\frac{\omega}{c}\right)^2 - k^2 = 0. \tag{1.32}$$

The solution for differential equation (1.31) have following standard form:

$$X(x) = A \sin k_x x + B \cos k_x x. \tag{1.33}$$

Now we can apply the boundary conditions for fixing the unknown coefficients. The boundary condition at $x = 0$ and $x = a$ requires that $\mathbf{B}_x$ (equation (1.26)) and $\frac{dX}{dx}$ vanishes. This results in

$$A = 0, \quad k_x = \frac{m\pi}{a} \quad m = 0, 1, 2, \dots . \tag{1.34}$$

Similarly, the differential equation for $Y(y)$ with corresponding boundary equations produces

$$k_y = \frac{n\pi}{b} \quad n = 0, 1, 2, \dots . \tag{1.35}$$

The final solution for the B_z component reads as follows:

$$\boxed{B_z = B_0 \cos(m\pi x/a) \cos(n\pi y/b)}. \tag{1.36}$$

This is the solution for the TE_{mn} mode. The wave number of the mode propagating in the waveguide can be expressed in terms of waveguide dimensions:

$$\boxed{k^2 = \left(\frac{\omega}{c}\right)^2 - \left(\left(\frac{m\pi}{a}\right)^2 + \left(\frac{n\pi}{b}\right)^2\right)} \tag{1.37}$$

Therefore, for proper guiding of the EM-wave, which corresponds to having real values of k, the frequency must be above the following cutoff frequency:

$$\omega_c = c\sqrt{\left(\frac{m\pi}{a}\right)^2 + \left(\frac{n\pi}{b}\right)^2} \tag{1.38}$$

Thus, the cutoff frequency of a waveguide is the lowest frequency at which a specific mode can propagate without attenuation. Below the cutoff frequency, the mode cannot propagate efficiently. Higher-order modes have higher cutoff frequencies, and the lowest cutoff frequency is associated with the fundamental mode. One can calculate the lowest cutoff frequency for a given waveguide for the mode TE_{10}:

$$\omega_{10} = c\pi/a. \tag{1.39}$$

Similarly, the lowest frequency, which can guide through is for the mode TM_{11},

$$\omega_{11} = c\sqrt{\left(\frac{\pi}{a}\right)^2 + \left(\frac{\pi}{b}\right)^2} \qquad (1.40)$$

One can observe from the cutoff frequency expression that the hollow conductor waveguides allow for the transmission of EM-wave in microwave frequencies. Optical fibers are used to transmit light waves with frequencies of a few hundred tera-hertz in the optical domain. They are cylindrical dielectric tubes of a very small diameter of 50–200 μm, which can guide optical signals to several hundreds of kilometers with minimal losses. Optical fiber can be considered as dielectric waveguides. The light propagation inside optical fiber can be simplified to the exact analytic solution of the Maxwell's equation for some special cases, as explained in the following sections.

1.4 Optical fiber and light propagation

Optical fibers are a type of waveguide used to transmit light signals over long distances with minimal loss and distortion. They are very widely used in telecommunications systems, optical sensors, and precision optical systems. Optical fibers have revolutionized telecommunications, providing the backbone for long-distance communication networks, internet connectivity, and high-speed data transmission. They are also widely used in medical imaging, sensing applications, and industrial systems. Continuous advancements in fiber optics technology have led to higher data transmission rates, improved fiber design, and expanded applications in various fields. Optical fibers offer several advantages over other communication mediums, such as copper wires or wireless transmission:

- High bandwidth: Optical fibers have a wide bandwidth, which stems from the high frequency and large number of modes they can support. Single-mode fibers, in particular, with their small core size, exhibit extremely high bandwidth potential. They can transmit light signals over a broad frequency range, covering wavelengths from the ultraviolet (UV) to the infrared (IR) spectrum.
- Low loss: Optical fibers are made from high-purity materials, typically silica glass or plastic polymers. These materials have low impurity levels, ensuring minimal absorption and scattering of light as it travels through the fiber. High-quality materials help maintain signal integrity and reduce losses. Also, the production of optical fibers involves precise manufacturing techniques to ensure uniformity and minimize defects. The core and cladding dimensions are carefully controlled, resulting in a well-defined waveguide structure that supports efficient light transmission with low losses. Therefore, optical fibers have low signal attenuation, enabling long-distance transmission with minimal loss.

– Immunity to electromagnetic interference: Since optical fibers transmit light signals, they are not affected by electromagnetic interference from nearby electrical sources.

Optical fibers consist of a cylindrically symmetric hair-thin high-quality transparent material with a few hundred-micrometer diameters. They allow low-loss transmission of light beams via total internal reflection. In principle, they can have various cross-sectional shapes with varying refractive index profiles along the cross-section. However, standard optical fibers, which are widely used for optical communication, are step-index fibers. They have a central core region with a diameter of a few microns and an outer cladding region with an overall fiber diameter of a few hundred microns. The fiber is typically further protected by a buffer layer, which provides mechanical strength and protection against external damage. On top of the buffer layer, a coating is applied for additional protection and to prevent moisture absorption (Figure 1.9). The core region is Germanium and doped to have a slightly higher refractive index compared to the cladding region (Figure 1.8). For a very small core diameter, typically one can have a single optical mode propagation inside the fiber. The principle of light propagation in optical fibers relies on the phenomenon of total internal reflection. When light enters the core of the fiber at an angle greater than the critical angle, it undergoes multiple internal reflections at the core-cladding interface. This ensures that the light remains

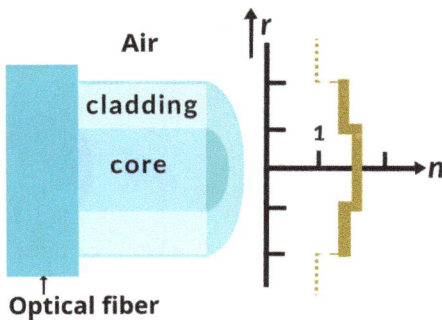

Figure 1.8: This figure shows the refractive index profile of a typical step-index single-mode fiber. The core has a slightly higher refractive index in comparison to the cladding due to a small Germanium doping.

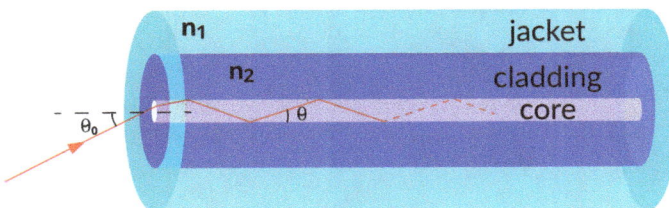

Figure 1.9: This figure shows a ray propagation scheme of light traveling through the fiber. The slightly lower refractive index of the cladding allows total internal reflection of light for a specific input incident angle on the fiber.

confined within the core and propagates along the length of the fiber with minimal attenuation. The light propagation inside a dielectric waveguide fiber can be analyzed by solving Maxwell's equations.

In the following sections, the mode analysis of the guided light in a fiber is described below, and single-mode fiber propagation conditions are explained.

1.4.1 EM-wave equation and its solution for an SM-fiber

The propagation of light within a step-index fiber can be analyzed by solving Maxwell's equations in cylindrical coordinates (Fig. 1.10) for isotropic dielectrics. The solutions for the core and cladding regions of the fiber are then matched to satisfy the boundary conditions at the core-cladding interface [2, 1, 6]. When considering the allowed modes inside the fiber, there are generally three scenarios: core-guided modes, cladding-guided modes, and radiation modes. As the name suggests, core-guided modes are allowed modes that are guided by the core-cladding interface and are mostly confined in the core region. On the other hand, cladding-guided modes are extended up to the cladding-buffer interface and guided by the combination of core-cladding and cladding-buffer interface. Radiation modes are unguided light that is not confined by the fiber. Our focus primarily lies on the natural modes (core-guided), which are guided modes near the core-cladding interface that decay exponentially in the transverse direction from the fiber's core. These modes are commonly referred to as core-guided modes. In the following, we first give a brief introduction to the guided modes in a step-index fiber and then analytically describe the field distribution.

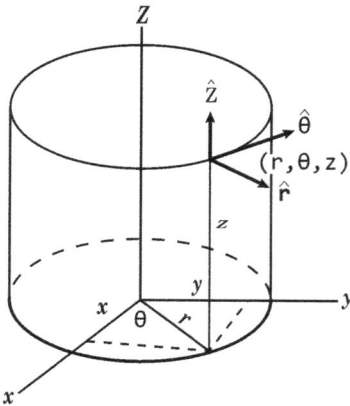

Figure 1.10: Cylindrical coordinate system for solving Maxwell's equations. The z-axis is chosen along the symmetry axis of the fiber.

Inside a step-index optical fiber, various core-guided modes can exist. These modes are characterized by their unique field distributions and propagation characteristics. Here are some of the common guided modes found in step-index fibers:

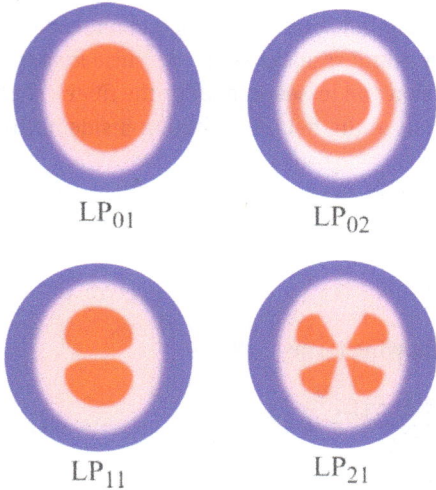

LP$_{01}$

LP$_{02}$

Figure 1.11: LP-modes, also known as linearly polarized modes, refer to the different electromagnetic field patterns that can propagate through an optical fiber. These modes are characterized by their spatial distribution of the electric field within the fiber cross-section. The LP-modes are classified based on their mode field distribution and the number of intensity peaks in the radial direction.

LP$_{11}$

LP$_{21}$

- Fundamental mode (LP$_{01}$): The fundamental mode, also known as LP$_{01}$ mode (Figure 1.11), is the lowest-order mode and has the highest power-carrying capability. It exhibits the simplest field distribution, with a single peak in the center of the fiber core. The LP$_{01}$ mode is often the most desirable mode for communication systems due to its low dispersion and minimal loss.
- Higher-order modes (LP$_{11}$, LP$_{21}$, etc.): Higher-order modes have additional peaks and nulls in their field distributions compared to the fundamental mode (Figure 1.11). These modes can support more than one wavelength or polarization state. However, higher-order modes generally have higher dispersion and higher attenuation than the fundamental mode, making them less desirable for long-distance communication.
- LP modes: LP (Linearly Polarized) modes are guided modes that have a specific polarization orientation along the fiber. They can be categorized into LP$_{11}$, LP$_{21}$, LP$_{02}$, LP$_{31}$, etc., depending on the number of intensity peaks and nulls in their field distributions. These modes have different propagation constants and may experience different levels of attenuation and dispersion.
- Higher radial-order modes (HE, EH Modes): In addition to LP modes, step-index fibers can also support higher radial-order modes, often referred to as HE (Higher-order Even) and EH (Higher-order Odd) modes. These modes have variations in both radial and azimuthal directions, resulting in more complex field distributions.

The number and types of guided modes supported by a step-index fiber depend on various factors, including the fiber's core size, refractive index profile, and operating wavelength. The mode structure and characteristics determine the fiber's overall performance in terms of dispersion, attenuation, and bandwidth. It is important to note that the actual modes supported by a specific step-index fiber will depend on its design parameters and operating conditions. Engineers and researchers study these modes to optimize fiber performance, minimize dispersion and losses, and tailor the fiber's characteristics to specific applications. We can analyze the field distribution inside a fiber from Maxwell's equations.

Inside an isotropic charge-free dielectric, where charge density (ρ) and current density (J) are zero,

$$\rho = 0, \quad J = 0, \tag{1.41}$$

the wave equation (equation (1.13)) has the following form:

$$\Delta E(r,t) - \frac{1}{v(r)^2} \frac{\partial^2 E(r,t)}{\partial t^2} = 0, \tag{1.42}$$

where Δ is the Laplacian operator and $v(r) = \frac{c}{n(r)}$ with c is the speed of light in vacuum and the refractive index of the medium is defined as $n(r) = \sqrt{\mu_r(r)\epsilon_r(r)}$. The wave equation for the H components is written by simply replacing E with H in the equation above.

Considering the cylindrical axis of symmetry of the fiber along the z-axis, the Laplace operator Δ in the cylindrical coordinates (r, ϕ, z), can be defined as follows:

$$\Delta = \partial_r{}^2 + \frac{1}{r^2}\partial_\phi{}^2 + \partial_z{}^2 \tag{1.43}$$

where $[\partial_r, \partial_\phi, \partial_z] = [\frac{\partial}{\partial r}, \frac{\partial}{\partial \phi}, \frac{\partial}{\partial z}]$.

The light propagates along the z-axis, which is the cylindrical symmetric axis of the fiber. Considering the angular frequency ω of light and the fiber propagation constant β, we can assume a time-dependent exponential form for the electric and magnetic field components:

$$\begin{bmatrix} E(r,t) \\ H(r,t) \end{bmatrix} = \begin{bmatrix} E(r,\phi) \\ H(r,\phi) \end{bmatrix} \exp(-i(\omega t + \beta z)). \tag{1.44}$$

As we are interested in the modes which are confined within the fiber, the propagation constant β has to be within a range of the propagation constants when the light travels exclusively in the cladding medium with refractive index n_2 and the propagation constant when light travels exclusively in the core medium with refractive index n_1:

$$n_2\left(\frac{\omega}{c}\right) < \beta < n_1\left(\frac{\omega}{c}\right). \tag{1.45}$$

Note that the propagation constant for light in vacuum is $k_0 = \frac{\omega}{c}$. Using Maxwell's equations (equation (1.8) and equation (1.10)), one can express the transversal components in terms of the axial field components E_z and H_z as follows:

$$E_r = \frac{i\beta}{\omega^2\mu\epsilon - \beta^2}\left(\partial_r E_z + \frac{\omega\mu}{\beta}\frac{\partial_\phi}{r}H_z\right) \tag{1.46}$$

$$E_\phi = \frac{i\beta}{\omega^2\mu\epsilon - \beta^2}\left(\frac{1}{r}\partial_\phi E_z - \frac{\omega\mu}{\beta}\frac{\partial_r}{r}H_z\right) \tag{1.47}$$

$$H_r = \frac{i\beta}{\omega^2\mu\epsilon - \beta^2}\left(\partial_r H_z - \frac{\omega\epsilon}{\beta}\frac{\partial_\phi}{r}E_z\right) \tag{1.48}$$

$$H_\phi = \frac{i\beta}{\omega^2\mu\epsilon - \beta^2}\left(\frac{1}{r}\partial_\phi H_z + \frac{\omega\epsilon}{\beta}\frac{\partial_r}{r}E_z\right) \tag{1.49}$$

Therefore, the wave equations can be solved only for the axial components (E_z and H_z) and the transversal components can be derived from the above equations.

The wave equation for the electromagnetic field component from equations (1.43) and (1.44) in cylindrical coordinates (Fig. 1.10) is as follows:

$$\left[\partial_r^2 + \frac{1}{r}\partial_r + \frac{1}{r^2}\partial_\phi^2 + (k^2 - \beta^2)\right]\begin{bmatrix}E_z\\H_z\end{bmatrix} = 0 \tag{1.50}$$

where $k = n\frac{\omega}{c}$ is the wave number for light traveling inside a medium with refractive index n. This standard differential equation can be separated into radial and azimuthal parts with the following ansatz:

$$\begin{bmatrix}E_z\\H_z\end{bmatrix} = \begin{bmatrix}E_0 e_z(r)\\H_0 h_z(r)\end{bmatrix}\exp(\pm il\phi) \tag{1.51}$$

Here, l is a positive integer, and $e_z(r)$, $h_z(r)$ are the electric and magnetic field functions defining the field configuration in the radial plane.

The differential equation arising from substituting equation (1.51) into the wave equation (equation (1.50)) takes the following form:

$$\left[\partial_r^2 + \frac{1}{r}\partial_r + \left(k^2 - \beta^2 - \frac{l^2}{r^2}\right)\right]\begin{bmatrix}e_z(r)\\h_z(r)\end{bmatrix} = 0 \tag{1.52}$$

This differential equation is well known (see Appendix B). Its solutions are known as the Bessel function or the modified Bessel function of order l. The order of the Bessel function depends on the coefficient of the last term of the differential equation.

When $h^2 = k^2 - \beta^2 > 0$, the solutions in terms of the Bessel functions of the first and second kind are as follows [8]:

$$\begin{bmatrix}e_z(r)\\h_z(r)\end{bmatrix} = \begin{bmatrix}c_1\\c_1^x\end{bmatrix}J_l(hr) + \begin{bmatrix}c_2\\c_2^x\end{bmatrix}Y_l(hr). \tag{1.53}$$

On the other hand, if $q^2 = k^2 - \beta^2 < 0$, the solutions in terms of the modified Bessel functions of the first and second kind are as follows:

$$\begin{bmatrix} e_z(r) \\ h_z(r) \end{bmatrix} = \begin{bmatrix} c_1 \\ c_1{}^x \end{bmatrix} I_l(qr) + \begin{bmatrix} c_2 \\ c_2{}^x \end{bmatrix} K_l(qr) \tag{1.54}$$

The complex coefficients c_1, $c_1{}^x$, c_2, $c_2{}^x$ are determined by the boundary conditions at the core-cladding interfaces.

For the core region: $r < a$, $k^2 - \beta^2 > 0$ and equation (1.53) are applicable. As the Bessel function of the second kind $Y_l(hr)$ diverges for $r \to 0$, therefore, to have meaningful modes at the central region of the fiber core, one needs to set the coefficient corresponding to this function equal to zero:

$$\begin{bmatrix} c_2 \\ c_2{}^x \end{bmatrix} = 0 \tag{1.55}$$

Consequently, the field equation in the core region $r < a$ is as follows:

$$\begin{bmatrix} e_z(r) \\ h_z(r) \end{bmatrix} = \begin{bmatrix} A \\ B \end{bmatrix} J_l(hr) \tag{1.56}$$

For cladding region: $r > a$, $\beta^2 - k^2 = 0$. Therefore, equation (1.54) is valid. However, except at the core-cladding interface, the electric field far away in the radial direction should vanish, i. e., as $r \to \infty$. As $J_l(hr)$ diverges in this region, one can set

$$\begin{bmatrix} c_1 \\ c_1{}^x \end{bmatrix} = 0 \tag{1.57}$$

The remaining modified Bessel function describes the field evolution in the cladding region, which means they represent the radially decaying evanescent field. Therefore, no power propagates along the radial direction. The electric field equation in the cladding region ($r > a$) is therefore as follows:

$$\begin{bmatrix} e_z(r) \\ h_z(r) \end{bmatrix} = \begin{bmatrix} C \\ D \end{bmatrix} K_l(qr) \tag{1.58}$$

Finding the expression for the core-guided mode propagating inside the fiber core involves estimating the complex coefficients A, B, C, D. This is done by satisfying the boundary conditions at the core-cladding interface such that the field components E_ϕ, E_z, H_ϕ, and H_z are continuous at the core-cladding interface. Substituting equations (1.56), (1.58) into the radial and azimuthal components of the fields in equations (1.46)–(1.49), one gets radial and azimuthal components in relation to z-components. For the continuity of the perpendicular field components at the core-cladding interface, a set of equations relating the four coefficients A, B, C, D, and propagation constant are obtained. For the existence of these nontrivial sets of equations, the determinant of the coefficients is set

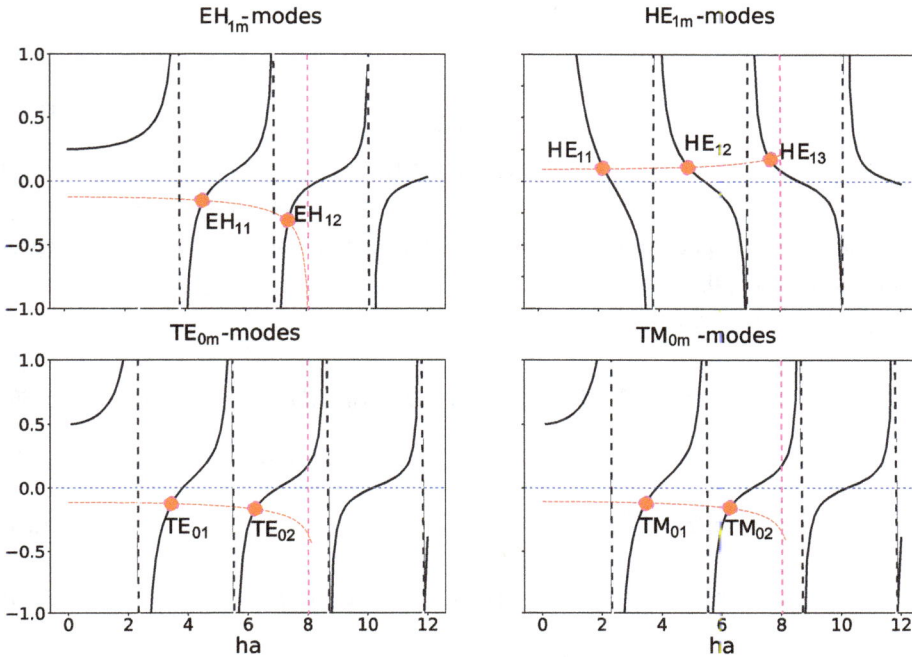

Figure 1.12: The solution to the transcendental equation for light propagation inside the fiber, i. e., equation (1.59). The intersection between the left- and right-hand side functions in the equation indicates the allowed propagation modes inside the fiber.

to zero. This generates the following relation, which allows finding the value of the propagation constant for a given set of values of l and ω,

$$\left[\frac{J'_l(ha)}{ha J_l(ha)} + \frac{K'_l(qa)}{qa K_l(qa)} \right]\left[n_1{}^2 \frac{J'_l(ha)}{ha J_l(ha)} + n_2{}^2 \frac{K'_l(qa)}{qa K_l(qa)} \right] = \left[\frac{1}{(qa)^2} + \frac{1}{(ha)^2} \right]^2 \left[\frac{l\beta}{k_0} \right]^2 \quad (1.59)$$

The relation between the field coefficients is as follows:

$$\frac{B}{A} = i\left(\frac{\beta s l}{\mu \omega} \right) \quad (1.60)$$

$$\frac{C}{A} = \zeta \quad (1.61)$$

$$\frac{D}{B} = \zeta \quad (1.62)$$

where

$$s = \left(\frac{1}{(qa)^2} + \frac{1}{(ha)^2} \right)\left(\frac{J'_l(ha)}{ha J_l(ha)} + \frac{K'_l(qa)}{qa K_l(qa)} \right)^{-1} \quad (1.63)$$

and

$$\zeta = \frac{J_l(ha)}{K_l(qa)} \tag{1.64}$$

$$n_2 k_0 \leq \beta \leq n_1 k_0 \tag{1.65}$$

The vector components of the field function inside the core and cladding are represented as follows [9].

For $r < a$ inside the fiber core,

$$e_r(r) = i\frac{q}{h\zeta}\left[(1 - sl)J_{l-1}(hr) - (1 + sl)J_{l+1}(hr)\right] \tag{1.66}$$

$$e_\phi(r) = -\frac{q}{h\zeta}\left[(1 - sl)J_{l-1}(hr) + (1 + sl)J_{l+1}(hr)\right] \tag{1.67}$$

$$e_z(r) = \frac{2q}{\zeta\beta}J_l(hr) \tag{1.68}$$

$$h_r(r) = \frac{\omega\epsilon_0 n_1^2 q}{h\beta\zeta}\left[(1 - s_1 l)J_{l-1}(hr) + (1 + s_1 l)J_{l+1}(hr)\right] \tag{1.69}$$

$$h_\phi(r) = i\frac{\omega\epsilon_0 n_1^2 q}{h\beta\zeta}\left[(1 - s_1 l)J_{l-1}(hr) - (1 + s_1 l)J_{l+1}(hr)\right] \tag{1.70}$$

$$h_z(r) = i\frac{2qsl}{\omega\mu\zeta}J_l(hr) \tag{1.71}$$

For $r > a$ in the cladding region,

$$e_r(r) = i\left[(1 - sl)K_{l-1}(qr) + (1 + sl)K_{l+1}(qr)\right] \tag{1.72}$$

$$e_\phi(r) = -\left[(1 - sl)K_{l-1}(qr) - (1 + sl)K_{l+1}(qr)\right] \tag{1.73}$$

$$e_z(r) = \frac{2q}{\beta}K_l(qr) \tag{1.74}$$

$$h_r(r) = \frac{\omega\epsilon_0 n_2^2}{\beta}\left[(1 - s_2 l)K_{l-1}(qr) - (1 + s_2 l)K_{l+1}(qr)\right] \tag{1.75}$$

$$h_\phi(r) = i\frac{\omega\epsilon_0 n_2^2}{\beta}\left[(1 - s_2 l)K_{l-1}(qr) + (1 + s_2 l)K_{l+1}(qr)\right] \tag{1.76}$$

$$h_z(r) = i\frac{2qsl}{\omega\mu}K_l(qr) \tag{1.77}$$

where $s_i = \frac{\beta^2}{k_0^2 n_i^2}s$, $i = 1, 2$.

$$J'(x) = \frac{\partial J(x)}{\partial x}, \quad K'(x) = \frac{\partial K(x)}{\partial x} \tag{1.78}$$

1.4.2 Propagation constant and fiber modes

When solving equation (1.59) numerically, a discrete set of values for the propagation constant β can be obtained, representing different propagation modes inside the fiber. The quadratic nature of equation (1.59) results in two sets of modes: HE-modes, for which $E_z > H_z$, and EH-modes, for which $E_z < H_z$. Depending on the value of l, there exist different solutions that characterize the various fiber propagating modes. As a result, the symbols EH_{lm} or HE_{lm} are used to denote these modes.

A special case of the general solution arises when $l = 0$, which is useful in defining modes in single-mode fibers:

- TM-modes correspond to the solutions of EH_{0m}. The characteristic mode equation for this case from equation (1.59) is

$$\frac{J_l(ha)}{haJ_0(ha)} = -\frac{n_2^2}{n_1^2} \frac{K_l(qa)}{qaK_0(qa)}.$$

(1.79)

- TE-modes correspond to the solutions of HE_{0m}. The characteristic mode equation for this case from equation (1.59) is

$$\frac{J_l(ha)}{haJ_0(ha)} = -\frac{n_2^2}{n_1^2} \frac{K_l(qa)}{qaK_0(qa)}.$$

(1.80)

The transverse magnetic (TM) and transverse electric (TE) modes have null magnetic and electric fields, respectively, in the direction of propagation. In other words, the field components are purely transverse, without any longitudinal components along the propagation direction.

Fundamental fiber parameter

The V-parameter, also known as the normalized frequency or V number, is a fundamental parameter used to quantify the relationship between the core size and the operating wavelength. It determines the number of guided modes in a fiber and plays a crucial role in fiber design and performance optimization. The V-parameter depends on the fiber properties, including the refractive indices of the core (n_1) and cladding (n_2), the core radius, and the wavelength of the light used (λ). It can be calculated using the equation

$$V = \frac{2\pi a}{\lambda} \sqrt{n_1^2 - n_2^2}$$

(1.81)

The V-parameter helps determine meaningful solutions of equation (1.59). By plotting the left and right sides of the equation, a graphical solution can be obtained. For example, when $l = 1$, the solutions are plotted in Figure 1.12, with the function plotted against the variable ha. The intersections of the curves represent the propagation modes in the fiber. There is a lower limit for V below, which a mode E_{lm} can propagate in the fiber. The cutoff value for the E_{lm} mode is given by the m roots of $J_l(ha) = 0$, such as $ha = 3.832, 7.016, 10.713$ for E_{11}, E_{12}, and E_{13} modes, respectively.

For HE-modes, the graphical solution can be obtained by plotting equation (1.59) similar to EH-modes. An interesting observation is that the HE_{11} mode has no cutoff, which means that this mode can always propagate. Therefore, HE_{11} mode is also referred to as the fundamental mode of the fiber.

1.4.2.1 Single-mode fiber condition

A single-mode (SM) optical fiber allows only a single transverse mode of light to propagate through. The mode allowed is the fundamental Gaussian mode due to the design properties of the fiber. In fiber optics, single-mode fibers are one of the most used fibers. The importance of SM-fibers is reflected by the 2009 Nobel Prize in Physics awarded to Charles K. Kao for his extensive theoretical contribution to SM-fibers. In the following, a very brief introduction to SM-mode propagation conditions in the fibers is introduced.

As discussed in the previous Section 1.4.2, while solving Maxwell's equations for the core-guided modes inside a fiber, one naturally obtains a very useful parameter. This parameter is called the "V-parameter" or the "normalized frequency" and depends on the guided wavelength, core radius, and the refractive index profile of the fiber. The V-parameter determines whether, for a given optical frequency, there is a single core-guided mode or multiple of them.

- For $V \leq 2.405$, there exists only a single core-guided mode and, therefore, a fiber for a given wavelength satisfying the above inequality (due to its refractive indices profile and radius) is called a single-mode fiber.
- For $V > 2.405$, more than one mode exists, which can be guided in the core and, therefore, fiber is called a multimode fiber.

Another way to visualize the single-mode fiber propagation condition is as follows. First, one can define the number of guided modes in terms of a normalized propagation constant b, which can be defined as follows:

$$b = \frac{a^2 w^2}{V^2} = \frac{(\beta/k)^2 - n_2^2}{n_1^2 - n_2^2}. \tag{1.82}$$

Now, from a plot similar to the one in Figure 1.12 for the equation above, normalized modes b versus V-parameter, one can find that there exists a minimum cutoff value above which a specific guided mode exists, except for HE_{11} mode which can exist for all values of the core diameter above zero. Therefore, the single-mode fiber condition exists for the fundamental mode below a certain V-parameter where only HE_{11} exists. The lowest nonfundamental mode inside the fiber is TM_{01}, the cutoff value of the parameter V for this then sets the lowest cutoff below which the single-mode propagation inside the fiber occurs. $V = 2.405$ is the lowest cut-off for this mode. Therefore, the single-mode fiber condition is $V < 2.405$.

1.5 Bottle resonator modes

Bottle microresonators are a specific type of optical microresonator that exhibit a unique geometric shape resembling a bottle or flask. These highly prolate shape dielectric structures have very special features for light propagation near their surface. The light spirals around the symmetry axis between two turning points, the so-called caustics (Figure 1.13), and cannot exit along the symmetry axis (z-axis) due to a kind of barrier in angular momentum space. While a comprehensive analysis of the light propagation in bottle resonators involves sophisticated mathematical and numerical techniques [8], we provide a brief overview here to highlight the concept and its relevance in the context of fiber-based resonators. In Chapter 3, we will delve deeper into the experimental aspects of fiber-based bottle resonators, including their fabrication and characterization.

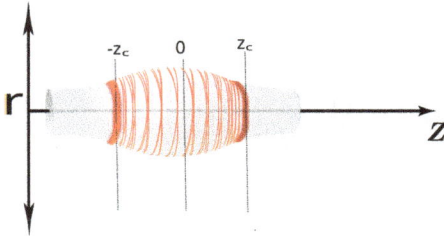

Figure 1.13: Light beam spiraling around a bottle resonator with parabolic profile along the axis. The maximum intensity is at the two caustic points along the axis. This figure is redrawn from [7] with permission, copyright (2005) by the American Physical Society.

From the Maxwell's equations, the electromagnetic fields (E, H) of the resonators satisfy the Helmholtz equation:

$$(\nabla^2 + k^2)\Psi = 0 \tag{1.83}$$

where Ψ represents the electromagnetic field vectors, $\Psi = \left[\begin{smallmatrix} E(r,t) \\ H(r,t) \end{smallmatrix} \right]$ and $|k| = n\,k_0 = n\frac{2\pi}{\lambda_0}$ is the wave number defined in terms of the wavelength of light in vacuum (λ_0) and the refractive index of the medium (n). Before solving equation (1.83), some approximations can be made based on the geometry of the bottle resonator:

- The profile of a bottle resonator is approximately a parabola along the symmetry axis (here z-axis):

$$R(z) = R_0\left(1 - \frac{1}{2}(\Delta k z)^2\right) \tag{1.84}$$

where R_0 is the central radius and Δk is the curvature of the resonator.
- The maximum angular momentum modes are close to the resonator surface as the light spirals around the bottle resonator. Assuming a small variation in resonator radius along the z-axis which is $\frac{dR}{dz} \ll 1$, the radial component of the wave vector

$k_r = (\frac{dR}{dZ})k_z$ is negligible compared to k_z and k_ϕ:

$$k = \sqrt{k_z{}^2 + k_\phi{}^2} \tag{1.85}$$

– Due to the cylindrical symmetry of the resonator, the projection of the angular momentum along the symmetry axis (z-axis) is a conserved quantity

$$\partial_z(k_\phi(z)R(z)) = 0 \tag{1.86}$$

– At the caustics, only the azimuthal components exists because the axial components vanish:

$$k_\phi(\pm z_c) = k\,; \quad k_z(\pm z_c) = 0 \tag{1.87}$$

Therefore, in the region, $-z_c \leq z \leq z_c$, the azimuthal and axial components of the wave vectors are

$$k_\phi(z) = k\,\frac{R_c}{R(z)}\,; \quad k_z(z) = \pm k\,\sqrt{1 - \left(\frac{R_c}{R(z)}\right)^2} \tag{1.88}$$

where $R_c = R(z_c)$.

Now coming back to equation (1.83), the cylindrical symmetry indicates the azimuthal part of the wave function to be of the form $\exp(im\phi)$ where m is the azimuthal quantum number. The adiabatic approximation allows the wave function to be written as a product of the axial wave function $Z(z)$ and the radial wave function $\Theta(r, R(z))$:

$$(\nabla^2 + k^2)\Theta(r, R(z))\,Z(z)\,e^{im\phi} = 0 \tag{1.89}$$

Equation (1.89) separates into two different equations:

$$\left(\partial_r{}^2 + \frac{1}{r}\partial_r + k_\phi{}^2 - \left(\frac{m}{r}\right)^2\right)\Theta(r, R(z)) = 0 \tag{1.90}$$

$$(\partial_z{}^2 + k^2 - k_\phi{}^2)Z(z) = 0 \tag{1.91}$$

Considering the radial equation (1.90), which is of the form of a Bessel differential equation, the solutions are the standard Bessel functions of the first and second kind, J_m and Y_m. Using the asymptotic behavior of the Bessel function and the condition of modes to be localized near the resonator surface, the radial mode function is the following: For $r \leq R(z)$,

$$\Theta(r, z) = \begin{bmatrix} A \\ B \end{bmatrix} J_m(k_\phi(z)r). \tag{1.92}$$

For $r > R(z)$,

$$\Theta(r,z) = \begin{bmatrix} C \\ D \end{bmatrix} Y_m(k_\phi(z)r). \tag{1.93}$$

The coefficients A, B, C, and D can be calculated using boundary conditions for the EM-field at a dielectric interface between two mediums. As the WGM resonator supports two polarization modes, which are TE- and TM-polarization, one can obtain relations between the coefficients at the dielectric interface and at $z = z_c$. The resonance condition for the two polarization modes can be written as follows.

For TM-modes,

$$\frac{n J'_m(k_0 n R_c)}{J_m(k_0 n R_c)} - \frac{n Y'_m(k_0 R_c)}{Y_m(k_0 R_c)} = 0. \tag{1.94}$$

For TE-modes,

$$\frac{J'_m(k_0 n R_c)}{J_m(k_0 n R_c)} - \frac{n Y'_m(k_0 R_c)}{Y_m(k_0 R_c)} = 0. \tag{1.95}$$

Axial equation can be recognized to be identical to the Harmonic oscillator differential equation. Using the relation $k_\phi(z) = \frac{m}{R(z)}$ and $R(z) = R_0\sqrt{1 + (\Delta k z)^2}$, the solution for equation (1.91) is

$$Z_{m,q} = \eta_{m,q} \cdot H_q\left(\sqrt{\frac{\Delta E_m}{2}} \cdot z\right) e^{-\frac{\Delta E_m}{4} z^2} \tag{1.96}$$

where H_q is the Hermite polynomial, $\eta_{m,q} = \left(\frac{\Delta E_m}{2^{2q+1}(q!)^2 \pi}\right)^{\frac{1}{4}}$ and

$$\Delta E_m = \frac{2 m \Delta k}{C_r R_0}, \tag{1.97}$$

where C_r is a correction factor in the case of deviation from the ideal parabolic profile. The parabolic profile of the bottle resonator creates a harmonic potential with discrete energy levels given by

$$E_{mq} = \left(q + \frac{1}{2}\right)\Delta E_m \tag{1.98}$$

where q is the axial quantum number. One can also derive the expression for the wave numbers, which are allowed

$$k_{m,q} = \sqrt{\frac{m^2}{(C_r R_0)^2} + \left(q + \frac{1}{2}\right)\frac{2 m \Delta k}{C_r R_0}}. \tag{1.99}$$

The intensity distribution of light along the resonator can be found from the wave function as follows [7, 5]:

$$I_{mq}(r,z) \propto |\Psi(r,z)|^2, \tag{1.100}$$

where $\Psi(r,z) = \Theta(r,z)Z_{m,q}(z)$.

Figure 1.14 shows the intensity profile, which is maximum at the two caustic points.

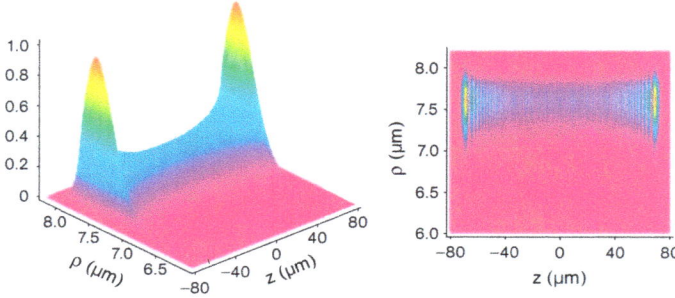

Figure 1.14: Light beam spiraling around a bottle resonator has intensity maximums at the two caustic points. This figure is reprinted from [7] with permission, copyright (2005) by the American Physical Society.

1.5.1 Free spectral range

The frequency separation for two consecutive axial (or azimuthal) modes is known as free spectral range (FSR) and this can be approximated as follows [7]. For small resonator ($\Delta k R_0 \ll 1$).

FSR along the axial direction is

$$\Delta \nu_m = \frac{c}{2\pi n}(k_{m+1,q} - k_{m,q}) \approx \frac{c}{2\pi n}\frac{1}{C_r R_0}. \tag{1.101}$$

FSR along the azimuthal direction is

$$\Delta \nu_m = \frac{c}{2\pi n}(k_{m+1,q} - k_{m,q}) \approx \frac{c}{2\pi n}\Delta k. \tag{1.102}$$

It is clear from the above expression that the azimuthal FSR of the resonator depends on the central radius (R_0), while the axial FSR is determined by the curvature (Δk) of the resonator.

Bibliography

[1] F. Bucholtz and J. M. Singley. A brief introduction to core modes in optical fiber. Technical report, Naval Research Laboratory, 2021.

[2] W. C. Chew. Lectures on theory of microwave and optical waveguides. arXiv preprint arXiv:2107.09672, 2021.

[3] D. J. Griffiths. *Introduction to electrodynamics*, 2005.

[4] L. Guilmette and et al. *The history of maxwell's equations*, 2012.

[5] C. Junge, S. Nickel, D. O'Shea, and A. Rauschenbeutel. Bottle microresonator with actively stabilized evanescent coupling. *Optics Letters*, 36(17):3488–3490, Sep 2011. https://doi.org/10.1364/OL.36.003488. URL https://opg.optica.org/ol/abstract.cfm?URI=ol-36-17-3488.

[6] F. L. Kien, J. Q. Liang, K. Hakuta, and V. I. Balykin. Field intensity distributions and polarization orientations in a vacuum-clad subwave length-diameter optical fiber. *Optics Communications*, 242(4–6):445–455, 2004.

[7] Y. Louyer, D. Meschede, and A. Rauschenbeutel. Tunable whispering-gallery-mode resonators for cavity quantum electrodynamics. *Physical Review A*, 72:031801, Sep 2005. https://doi.org/10.1103/PhysRevA.72.031801. URL https://link.aps.org/doi/10.1103/PhysRevA.72.031801.

[8] M. Pöllinger. *Bottle microresonators for applications in quantum optics and all-optical signal processing*. PhD thesis, Mainz, Univ., Diss., 2011, 2010.

[9] C. Wuttke. *Thermal excitations of optical nanofibers measured with a fiber-integrated Fabry-Pérot cavity*. PhD thesis, Mainz, Univ., Diss., 2014, 2014.

2 Optical resonators

Light propagation in the presence of boundary conditions in an optical system can have a profound impact on the phase and frequency responses of the optical device. Optical resonators are an excellent example of this phenomenon. They consist of two or more highly reflecting surfaces. The boundary conditions set by these surfaces allow light to be stored and transmitted only at certain discrete frequencies, known as resonance frequencies, and only specific intensity patterns are allowed, known as cavity modes. The boundary conditions, which mean the mirror´s reflectivity, distance, and curvature, determine the optical characteristics of the cavity, e. g., the time interval of light storage, the width of the transmission spectrum and the power gain.

The power gain at resonance and the suppression (filtering) of nonresonant light open up tremendous applications of optical resonators in various domains of experimental physics. These applications span multiple fields of experimental science and engineering. This chapter serves as a concise introduction to the fundamentals of cavities, laying the groundwork for a comprehensive exploration of fiber-based resonators in the next chapter. In the following sections, we will discuss how the cavity configuration and the properties of the mirrors determine the power gain or suppression of light due to interference effects. The chapter includes a concise introduction to the fundamental concepts of Fabry–Perot resonators and their practical applications. We delve into the diverse applications of optical resonators, showcasing a specific example of their usage in a gravitational wave detector. The methodologies harnessed in this specific application reverberate throughout the realm of optical resonators. The central aim of this exploration is to elucidate the fundamental tenets and innovative concepts that underlie the attainment of unparalleled levels of precision using optical resonators. Using the gravitational wave detection example, we see the role of optical resonators in achieving a significant milestone in precision measurements.

2.1 Fabry–Perot cavity

A Fabry–Perot cavity is a specific type of optical resonator that consists of two highly reflective mirrors positioned parallel to each other Figure 2.1. These mirrors create a confined space where light can be trapped and undergo multiple reflections. The name "Fabry–Perot" comes from the two French physicists, Charles Fabry and Alfred Perot, who first studied and characterized this type of resonator. The behavior of light within a Fabry–Perot cavity is governed by the interference of the reflected light waves between the mirrors. When the optical path length between the mirrors is an integer multiple of the wavelength of the light, constructive interference occurs, resulting in the enhancement of light intensity at certain frequencies or wavelengths. This phenomenon is known as resonance, and the corresponding frequencies are called resonance frequencies. Fabry–Perot cavity also has highly frequency selective transmission of light, which

https://doi.org/10.1515/9783110636260-002

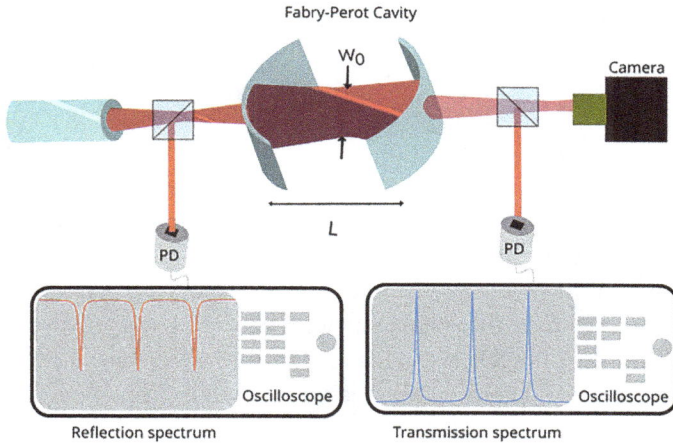

Figure 2.1: Depiction of a Fabry–Perot cavity (up) with a typical reflection and transmission spectrum (down). Here, PD indicates photodiode and ω_o is the waist of the light beam inside the cavity (length = L). A camera is used to observe the spatial modes of the cavity in the transmission port. Red and blue curves show the reflection and the transmission signals observed while scanning the laser frequency.

means sharp resonance peaks at certain input light frequencies. At these specific resonance frequencies, the transmission results from constructive interference among the light beams leaking through the output mirror after each round trip. On the input side, an intensity reduction results from the destructive interference between the directly reflected beam and the light leaking from the cavity through the input mirror after each round trip. The length, reflectivity, and alignment of the mirrors in a Fabry–Perot cavity determine its optical characteristics, such as the finesse (a measure of its ability to discriminate between different wavelengths) and the free spectral range (the spacing between consecutive resonance frequencies). These properties can be finely tuned and optimized for specific applications by adjusting the cavity length and mirror reflectivities. Quantitative analysis is presented in the following sections.

2.1.1 Confining light between reflecting surfaces

Consider a cavity arrangement of two perfectly reflecting mirrors facing each other with a distance l between them. For a light beam traveling between the mirrors, when the round-trip phase accumulated is an integer multiple of 2π, there is constructive interference at the mirrors. This condition allows light with a wavelength λ_n to be confined between the mirrors:

$$\Delta\phi = \left(\frac{2\pi}{\lambda_n}\right)2l = 2n\pi. \tag{2.1}$$

The constructive interference condition reinforces a wavelength-selective condition for trapped light. We can express it in terms of the corresponding frequencies:

$$v_n = \left(\frac{c}{\lambda_n} \right) = n \left(\frac{c}{2l} \right). \tag{2.2}$$

Thus, for a fixed mirror separation, only specific frequencies or longitudinal modes of light can exist within the cavity. These discrete wavelength values define the particular standing wave patterns, which can be confined inside the cavity. Therefore, the longitudinal modes of a cavity correspond to those stable standing wave patterns, which are constructively reinforced between the two bounding mirrors. The frequency interval between these longitudinal modes is known as the free spectral range (FSR) of the resonator, as written below:

$$\Delta v_{FSR} = v_{n+1} - v_n = n \frac{c}{2l}. \tag{2.3}$$

As evident from the aforementioned expression, the free spectral range (FSR) of an optical resonator is exclusively dictated by its geometric properties, specifically the length of the cavity. The relationship is inversely proportional, meaning that as the length of the resonator increases, the FSR decreases. Consequently, longer resonators have smaller frequency spacings between their longitudinal modes. This observation highlights the influence of the resonator's physical dimensions on its spectral characteristics.

2.1.2 Linewidth of a Fabry–Perot cavity

In a Fabry–Perot cavity, the behavior of light can be significantly influenced by the reflective properties of the mirrors. An ideal Fabry–Perot cavity would exhibit a characteristic pattern of delta resonance peaks. These resonance peaks are very narrow and have a distinct shape resembling delta functions. However, it is important to note that in practice, real-world factors such as mirror imperfections, optical losses, and environmental conditions can broaden and distort the resonance peaks.

The real mirrors used in practical Fabry–Perot cavities have finite reflectivity. They do not reflect all incident light perfectly but instead have some degree of loss. This finite reflectivity leads to a broadening of the resonance peaks (Figure 2.2). The broader bandwidth (linewidth) of transmitted or reflected light at the resonance frequencies can affect the spectral selectivity and overall performance of the cavity. In the following, we will calculate the transmission and reflection spectrum of a Fabry–Perot cavity with finite reflectivity and define the important parameters such as finesse and linewidth of a cavity.

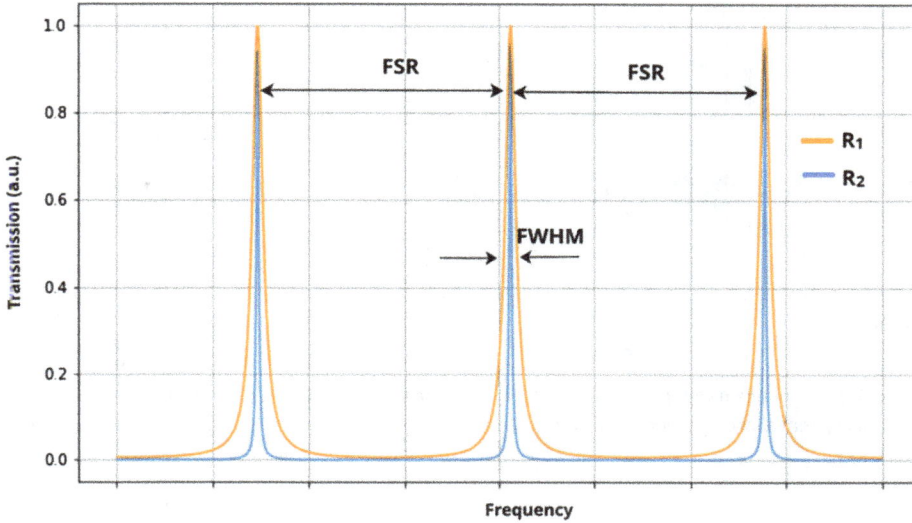

Figure 2.2: Transmission profile of a Fabry–Perot cavity demonstrating the frequency selective transmission response. The periodic cavity resonances are separated in frequency by a free-spectral-range of the cavity. R_1 and R_2 are two arbitrarily chosen mirror reflectivities to illustrate the effect on the cavity linewidth.

2.1.3 Plane-wave analysis

When a light beam is injected into an optical resonator, it interacts with the reflecting surface of either of the two mirrors. This interaction can be approximated in the context of a plane wave model with simple geometrical optics. The light beam incident on the reflecting surface encounters a boundary between two different mediums, Figure 2.3. The laws of reflection and refraction govern the path of the beam as it traverses between these mediums. To understand the behavior of light inside a Fabry–Perot cavity, an analytical expression for the transmission and reflection functions can be derived. In Figure 2.3, we consider a light beam with an amplitude of E_0 incident on the partially reflecting surface, which has reflectivity and transmission amplitudes of r_1 and t_1, respectively. Similarly, the end mirror (or the output mirror) has reflectivity and transmission amplitudes of r_2 and t_2, respectively. Note that the more convenient and directly measurable parameters are the power reflectivity ($R_i = r_i^2$), and transmitivity ($T_i = t_i^2$) of the cavity mirrors.

The effective amplitude E_c for the light circulating inside the cavity is composed of the transmitted amplitude through the input mirror, as well as successive terms resulting from multiple reflections between the mirrors. These terms are scaled by the reflec-

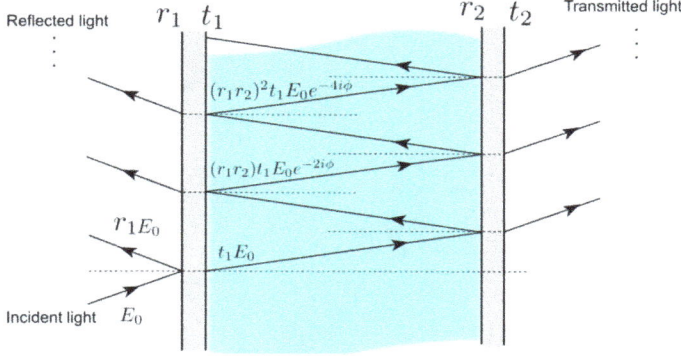

Figure 2.3: A light beam with plane wavefront is incident on two infinitely extended reflecting surfaces. This ideal system allows us to ignore the effects that can arise from the diffraction for the finite-size reflectors.

tivity functions of the mirrors and a propagation phase, equation (2.5). For a plane wave,

$$E = E_0 \exp i(\omega t - \mathbf{k}.\mathbf{r}) \tag{2.4}$$

where ω is the frequency, $\mathbf{k}$ is the propagation vector with $|\mathbf{k}| = \frac{2\pi}{\lambda}$, circulation electric field amplitude can easily be written as follows:

$$\begin{aligned}
E_c &= t_1 E_0 + t_1 E_0\, r_1 r_2 e^{-i2\phi} + t_1 E_0 \left(r_1 r_2 e^{-i2\phi}\right)^2 + \cdots. \\
&= t_1 E_0 \sum_{n=0}^{\infty} \left(r_1 r_2 e^{-i2\phi}\right)^n \\
&= \frac{t_1 E_0}{1 - r_1 r_2 e^{-i2\phi}}
\end{aligned} \tag{2.5}$$

Here, the phase factor 2ϕ includes the round trip propagation phase, $\phi = \omega L/c = \pi v/\Delta v_{\mathrm{FSR}}$, and any additional phase after reflection on the mirrors. Constructive interference between the multiple reflections between the mirrors enhances the power and signal within the resonator. One can calculate power gain factor inside the cavity as follows:

$$G_{\mathrm{FP}} = \left|\frac{E_c}{E_0}\right|^2 = \frac{T_1}{(1 - \sqrt{R_1}\sqrt{R_2})^2 + 4\sqrt{R_1}\sqrt{R_2}\sin^2\phi} \tag{2.6}$$

The transmission amplitude through the output mirror of the cavity, denoted as E_T, can be calculated from the circulating amplitude E_c by incorporating the transmission coefficient t_2 and a phase factor $e^{-i\phi}$:

$$E_T = t_2 E_c\, e^{-i\phi} = E_0 \frac{t_1 t_2 e^{-i\phi}}{1 - r_1 r_2 e^{-i2\phi}} \tag{2.7}$$

The transmission function of a Fabry–Perot cavity can be expressed as follows:

$$T_{\text{FP}} = \left| \frac{E_T}{E_0} \right|^2 = \frac{T^2}{|1 - Re^{i\phi}|^2} \tag{2.8}$$

Here, we define $T = t_1 t_2$ and $R = r_1 r_2$, assuming no internal losses at the interfaces, $t_i^2 + r_i^2 = 1$ where $i \in \{1, 2\}$.

Therefore, the transmitted intensity through an FP-cavity can be written as

$$I_T = I_0 \frac{1}{(1 - R)^2 + 4R \sin^2(\frac{\phi}{2})} = \frac{T_m}{1 + (\frac{2F}{\pi})^2 \sin^2(\frac{\phi}{2})} \tag{2.9}$$

where $T_m = \frac{I_0}{(1-R)^2}$ and $F = \frac{\pi \sqrt{R}}{(1-R)}$.

T_m is defined as the maximum transmission function and F is known as the finesse of the cavity.

Finesse

Finesse F is an important parameter of an optical resonator. It specifies the number of round trips a photon makes inside the optical resonator before it is lost either by leaking out of the cavity or gets lost in case of losses in the boundary surfaces. Finesse is a measure of the resonator's ability to enhance the constructive interference of light within the cavity. It is also defined as the ratio of the free spectral range (FSR) to the full width at half-maximum (FWHM) of a resonance peak in the resonator's transmission spectrum.

Cavity linewidth

The linewidth of a resonator is defined as the frequency interval between two points at half of the maximum intensity. In the above equation (equation (2.9)), one can find frequency values where $I = I_m/2$ (I_m is the maximum transmission intensity):

$$\sin^2\left(\frac{\pi \nu}{\Delta \nu_{\text{FSR}}} \right) = \left(\frac{\pi}{2F} \right)^2 \tag{2.10}$$

$$\nu = \pm\left(\frac{\Delta \nu_{\text{FSR}}}{\pi} \right) \sin^{-1}\left(\frac{\pi}{2F} \right) \tag{2.11}$$

For large finesse values, $\sin^{-1}(\frac{\pi}{2F}) \sim \frac{\pi}{2F}$, therefore,

$$\nu = \pm\left(\frac{\Delta \nu_{\text{FSR}}}{2F} \right) \tag{2.12}$$

The linewidth of the cavity is

$$\Delta \nu_{\text{FWHM}} = \left(\frac{\Delta \nu_{\text{FSR}}}{F} \right). \tag{2.13}$$

Quality factor

The quality factor (Q-factor) of a resonator is a dimensionless parameter that characterizes the efficiency and performance of the resonator in maintaining a stable and sharp resonance. It is a measure of how well a resonator can store and transfer energy before it dissipates due to losses. The higher the Q-factor, the more energy can be stored in the resonator and the narrower the resonance peak.

Mathematically, the Q-factor is defined as the ratio of the resonant frequency ν_0 to the full-width at half-maximum (FWHM) of the resonance curve $\Delta\nu_{FWHM}$:

$$Q = \frac{\nu_0}{\Delta\nu_{FWHM}} \tag{2.14}$$

Reflectivity function

The reflectivity function of the cavity is calculated from the interference amplitude of the light directly reflected from the input mirror and the circulated field leaking out of the cavity from the input mirror. Similar to equation (2.7), the total reflected electric field amplitude is

$$E_R = -r_1 E_0 + E_0 \frac{t_1^2 r_2 e^{-i2\phi}}{1 - r_1 r_2 e^{-i2\phi}} \tag{2.15}$$

The reflectivity function can be obtained from the above expression

$$R_{FP} = \left|\frac{E_R}{E_0}\right|^2 = \frac{(\sqrt{R_1} - \sqrt{R_2})^2 + 4\sqrt{R_1}\sqrt{R_2}\sin^2\phi}{(1 - \sqrt{R_1}\sqrt{R_2})^2 + 4\sqrt{R_1}\sqrt{R_2}\sin^2\phi}. \tag{2.16}$$

Here, we have used $t_1^2 = 1 - r_1^2$ and $R_{1,2} = r_{1,2}^2$.

2.1.3.1 Finesse of a cavity with losses

In the previous analysis of the cavity transfer function, we have not considered any losses. These losses arise from either—absorption, scattering, or clipping losses from the mirrors—or due to the attenuation from the medium between the cavity mirrors. In a realistic scenario, these losses contribute to a varying degree depending upon their absolute values.

The expression for the transmission, and reflection amplitudes of equation (2.7) and equation (2.15) change to the following form:

$$E_T = E_0 \frac{t_1 t_2}{1 - r_1 r_2 e^{-l_c - i2\phi}} \tag{2.17}$$

$$E_R = -r_1 E_0 + E_0 \frac{t_1^2 r_2 e^{-l_c - i2\phi}}{1 - r_1 r_2 e^{-l_c - i2\phi}} \tag{2.18}$$

There is just an extra exponential term e^{-l_c} along with the propagation phase. This exponential loss term accounts for the attenuation of the light beam each time traversing between the two mirrors. l_c accounts for any losses at the mirrors (say L_1 and L_2) or the attenuation in the medium between the cavity al where a is the attenuation constant. The power transmission and reflection function of the cavity with losses can be calculated from the above expressions. The expression contains a finesse parameter, a function of mirror reflectivity and losses [15]:

$$f = \frac{4\sqrt{R_1 R_2}e^{-l_c}}{1 - \sqrt{R_1 R_2}e^{-l_c}} \tag{2.19}$$

Intracavity power gain function:

$$G_{\text{FP}} = \frac{T_1}{(1 - \sqrt{R_1 R_2}e^{-l_c})^2(1 + f\sin^2(\phi))} \tag{2.20}$$

Power transmission function:

$$T_{\text{FP}} = \frac{T_1 T_2}{(1 - \sqrt{R_1 R_2}e^{-l_c})^2(1 + f\sin^2(\phi))} \tag{2.21}$$

Power reflection function:

$$R_{\text{FP}} = \frac{R_1 + (1 - L_1)^2 R_2 e^{-l_c} + 2T_1\sqrt{R_1 R_2}e^{-l_c}\cos\phi}{(1 - \sqrt{R_1 R_2}e^{-l_c})^2(1 + f\sin^2(\phi))} \tag{2.22}$$

Historically, analysis of cavity transmission included a description in terms of loss and gain function for the laser power circulating inside the cavity. This notation helps construct generalized expressions for any cavity with losses and gains within the cavity. In this notation, the mirror reflectivities are

$$R_1 = e^{-T_1}, \quad R_2 = e^{-T_2} \quad \text{where } T_1 = t_1^2, \quad T_2 = t_2^2. \tag{2.23}$$

In this representation, the cavity transmission function of equation (2.7) as follows:

$$\frac{E_T}{E_0} = \frac{t_1 t_2}{\sqrt{r_1 r_2}}\left(\frac{\sqrt{g}}{1 - g}\right) \tag{2.24}$$

where g is defined as the gain factor for the light inside the cavity. This terminology comes from the analysis of laser cavity, where the actual gain medium is placed between the circulating beam [15]. For the passive cavity,

$$g = r_1 r_2 e^{-l_c} \quad \text{or} \quad g = e^{-(l_c + T_1 + T_2)} \tag{2.25}$$

Now the finesse of the cavity introduced before $F = \frac{\pi\sqrt{R}}{(1-R)}$ can be written in a generalized form as

$$F = \frac{\pi \sqrt{g}}{1 - g} \sim \frac{2\pi}{L_1 + L_2 + T_1 + T_2} \tag{2.26}$$

For a high finesse cavity, this is also a ratio between two frequency parameters for an optical resonator. These are the free spectral range (FSR) and the linewidth of the resonator:

$$F = \frac{\Delta \nu_{\text{FSR}}}{\Delta \nu_{\text{FWHM}}} \tag{2.27}$$

This can be obtained directly from the linewidth definition of the resonator, equation (2.13).

2.1.4 Spherical cavity and Gaussian modes

In the previous section, the Fabry–Perot resonator analysis considered the incident light on the cavity as plane waves that extend infinitely on each axis perpendicular to the propagation direction. Likewise, the cavity mirrors were assumed to be infinite planes located at a distance equal to the cavity length. This simplified approach allows us to neglect diffraction losses and to treat the system using ray optics, which provides an idealized view of the light spectrum within the cavity.

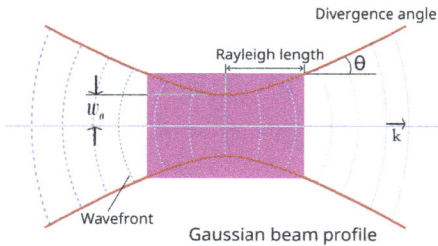

Figure 2.4: Spatial profile of a laser beam as a Gaussian beam.

However, one of the most commonly used light sources for a more realistic scenario is a laser beam, Figure 2.4. The spatial profile and the description of the laser beams are mathematically expressed by Hermite–Gaussian modes, which are a set of solutions to the wave equation:

$$(\nabla^2 + k^2)\mathbf{E} = 0. \tag{2.28}$$

The Hermite–Gaussian solutions are obtained by solving the above time-independent wave equation under the paraxial approximation, which assumes that the beam is nearly collimated and propagates in the paraxial regime. A typical solution presents the following time-independent spatial profile of the beam:

First Nine Transverse Mode Functions

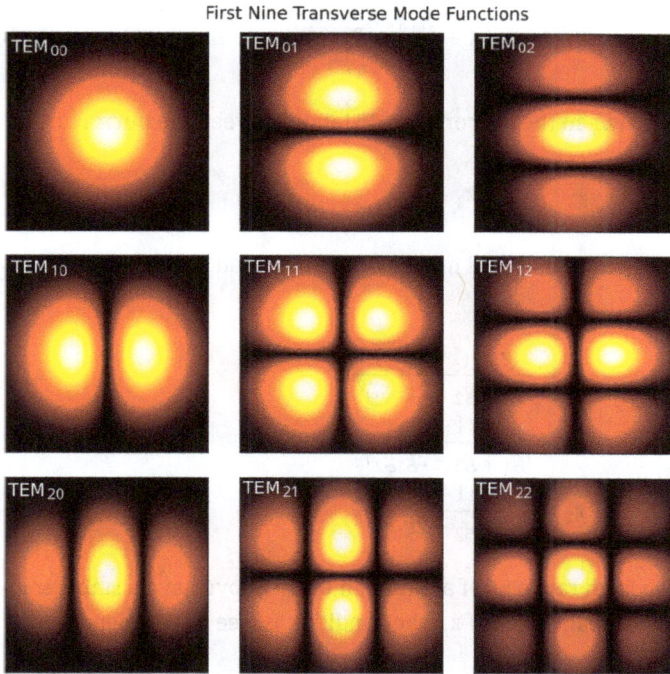

Figure 2.5: The transverse mode of a cavity demonstrates certain spatial patterns, which are represented by Hermite–Gaussian modes.

2.1.6 Stability parameter

The stability parameter is a concept used to analyze the ability of a resonator to keep the beam confined after many round trips once a light is launched parallel to the cavity axis. This concept is particularly useful for resonators that involves curved mirrors, such as Fabry–Perot cavities and other laser resonators. It helps determine whether the resonator will maintain a stable optical mode configuration or if the beam will diverge or become unstable.

To confine a Gaussian laser beam within an optical resonator, it is necessary to achieve perfect overlap between the radii of curvature of the cavity mirrors and the wavefront of the beam at the mirror positions. Let us consider a cavity configuration with mirrors having radii of curvature R_1 and R_2 at positions z_1 and z_2, respectively. The distance between the mirrors is denoted by l. For constructive reinforcement of the beam within the cavity, we can express the following conditions:

$$R(z_1) = z_1\left(1 + \frac{z_R^2}{z_1^2}\right) = -R_1 \qquad (2.36)$$

$$R(z_2) = z_1\left(1 + \frac{z_R^2}{z_2^2}\right) = R_2 \tag{2.37}$$

with a reference point in between the mirrors, the distance between the mirrors is

$$l = z_2 - z_1 \tag{2.38}$$

For above equations, expression for z_R and mirror locations z_1 and z_2 can be written as follows:

$$z_R^2 = \frac{g_1 g_2 (1 - g_1 g_2)}{(g_1 + g_2 - 2g_1 g_2)^2} l^2 \tag{2.39}$$

$$z_1 = \frac{g_2(1 - g_1)}{(g_1 + g_2 - 2g_1 g_2)} l \tag{2.40}$$

$$z_2 = \frac{g_1(1 - g_2)}{(g_1 + g_2 - 2g_1 g_2)} l \tag{2.41}$$

The expression for the Rayleigh range z_R of a Gaussian beam above have a stable solution for real values of z_R. For real values of z_R, the condition is (see also Appendix A):

$$0 \leq g_1 g_2 \leq 1 \tag{2.42}$$

Therefore, the stability condition for an optical resonator determines whether the resonator configuration is stable or not. The stability is characterized by the stability parameters, g_1 and g_2, which are derived from the mirror radii of curvature (R_1 and R_2) and the distance between the mirrors (l). In all the expressions above, g_1 and g_2 are known as the stability parameters. They are functions of the distance between the mirrors and the radius of curvature of the cavity

$$g_1 = 1 - \frac{l}{R_1} \tag{2.43}$$

$$g_2 = 1 - \frac{l}{R_2} \tag{2.44}$$

Based on the above expression, it is straightforward to know whether a cavity mirror configuration is stable (Figure 2.6).

One can express cavity waist and the waist of the beam on the two cavity mirrors as follows:

- Waist of the Gaussian beam

$$w_0^2 = \frac{l\lambda}{\pi}\left(\frac{g_1 g_2 (1 - g_1 g_2)}{(g_1 + g_2 - 2g_1 g_2)^2}\right)^{\frac{1}{2}} \tag{2.45}$$

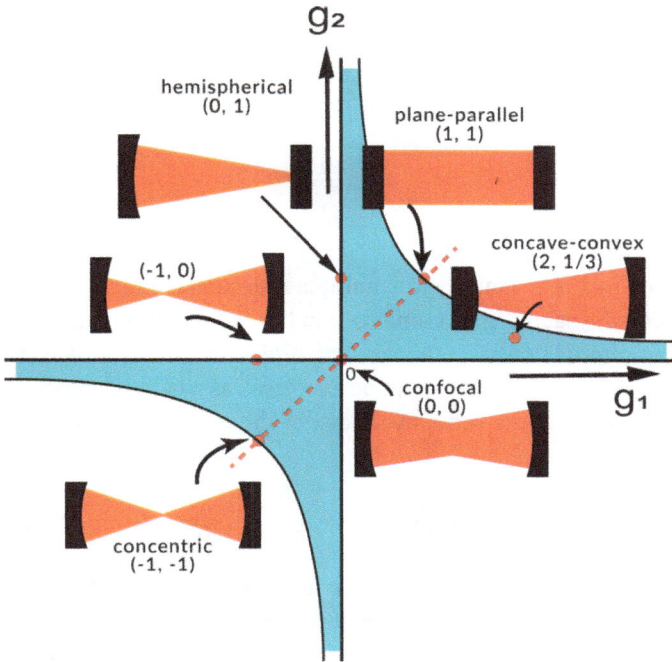

Figure 2.6: Stability regions for a Fabry–Perot cavity with different geometries.

– Waist of the beam at mirrors

$$w_1^2 = \frac{l\lambda}{\pi}\left(\frac{g_2}{g_1(1 - g_1 g_2)^{\frac{1}{2}}}\right)^{\frac{1}{2}} \tag{2.46}$$

$$w_2^2 = \frac{l\lambda}{\pi}\left(\frac{g_1}{g_2(1 - g_1 g_2)^{\frac{1}{2}}}\right)^{\frac{1}{2}} \tag{2.47}$$

2.1.7 Half-symmetric cavity

If the optical resonator consists of a concave and a plane mirror, then

$$R_1 = \infty \quad \text{and} \quad g_1 = 1, \, g_2 = 1 - \frac{l}{R_2} \tag{2.48}$$

Beam Waist, which lies on the plane mirror is given by

$$w_0^2 = \frac{l\lambda}{\pi}\left(\frac{g_2}{(1 - g_2)^{\frac{1}{2}}}\right)^{\frac{1}{2}}$$

The waist of the beam on the concave mirror is given by

$$w_2^2 = \frac{l\lambda}{\pi}\left(\frac{1}{g_2(1-g_2)}\right)^{\frac{1}{2}}$$

(2.49)

2.2 Cavity stabilization

In section 2.1, we have seen that at resonance the multiple reflections of the light constructively interferes inside the cavity. The resonance condition is achieved when the cavity length is an integer multiple of half the wavelength of light. This results in an expression for the resonance frequency, which can be expressed as a periodic function dependent on the cavity length and the refractive index of the medium between the mirrors:

$$\nu_N = N\frac{c_0}{2nL}$$

(2.50)

c_0 being the speed of light in vacuum, n is the refractive index of the medium between the mirrors, N is a positive integer, and L is the length of the cavity. To ensure stable resonance frequencies, it is important to address fluctuations in the cavity length and the refractive index. We want to estimate the relative frequency fluctuations arising from the cavity length and the refractive index changes in the medium.

The relative frequency stability $\frac{\Delta\nu}{\nu}$ or the relative resonance frequency error can be expressed as:

$$\frac{\Delta\nu}{\nu} = \sqrt{\left(\frac{\Delta L}{L}\right)^2 + \left(\frac{\Delta n}{n}\right)^2},$$

(2.51)

where ν is the cavity resonance frequency.

Therefore, changes in the cavity length or refractive index can directly impact the resonance frequency of an optical cavity. These changes can occur due to external factors such as mechanical deformation, thermal expansion of the cavity material, or variations in the refractive index caused by temperature and pressure.

To stabilize the resonance condition of the cavity, two approaches can be employed. The first approach involves constructing optical cavities with exceptionally high passive length stability. This means designing the cavity and its components in a way that minimizes the effects of external factors on the cavity length and refractive index. By using stable materials, vacuum and thermal insulation, and active temperature stabilization, the cavity can maintain its resonance condition over a wide range of operating conditions. The second approach is to utilize a tunable cavity, where the cavity length can be actively adjusted to maintain the resonance condition at a specific frequency. This active stabilization method involves continuously monitoring the cavity length and making real-time adjustments using feedback control systems. By employing sensors, actuators, and control algorithms, the cavity length can be actively controlled to counteract any changes caused by external factors. This ensures that the resonance condition remains stable and aligned with the desired frequency.

2.2.1 Ultrastable cavities

Ultrastable cavities, as the name suggests, are renowned for their exceptional frequency stability and serve as highly reliable devices for frequency referencing. These cavities are meticulously engineered using advanced materials and precise mechanics to achieve an exceptionally high level of passive stability. The level of technology involved in ultrastable cavities can be intuitively appreciated by the fact that the frequency noise in these cavities is limited to the thermal noise of the coating material for the best ultrastable cavities. This remarkably means that for such a system, the frequency noise is dominated by the Brownian motion of the molecules in the coating [17, 7, 10], and all other mechanical and electrical noises are suppressed much below the coating noise.

Typically, "ultrastable cavity" term is used to address a system that has its relative length stability $\Delta L/L = \Delta v/v \leq 10^{-14}$. As the length or the frequency drifts cannot be removed completely for all integration times, the cavity stability is defined for a specific integration time. The best-reported cavity stability is 6.5×10^{-17} at an integration time between 0.8–80 seconds for a silicon cavity [14], published in the year 2019. The cavity was under cryogenically cooled conditions at 4 K.

Constructing an ultrastable cavity necessitates a thorough assessment of the impact of various external factors on frequency and length stability. The following considerations are crucial in achieving high passive stability:

-   *Temperature variations*: Temperature fluctuations can change the effective cavity length because of the thermal expansion in the cavity mirror mounts. To achieve exceptional thermal stability in state-of-the-art ultrastable cavities, significant efforts are made to minimize the impact of temperature fluctuations on cavity frequency noise, reducing it to a level significantly below the frequency noise from mirror coatings. One key factor in reducing alignment sensitivity is to create a monolithic system of the two cavity mirrors with a low-expansion spacer material. In ultrastable cavities, the two cavity mirrors are optically contacted directly onto a glass spacer. "Optical contacting" here means a bonding method between highly polished surfaces utilizing intermolecular adhesion. Therefore, both the mirror surfaces and the surface of the spacer are super polished with the surface flatness of $\sim \frac{\lambda}{10}$. The optical contacting method produces a single monolithic system of mirrors and a spacer resulting in an alignment-free cavity. Therefore, the cavity is inherently stable. Furthermore, the temperature sensitivity of the cavity can be drastically reduced by using a spacer made out of special glass (or another material) with a very small thermal expansion coefficient:

$$\Delta L = \alpha L \Delta T \tag{2.52}$$

α is the linear thermal expansion coefficient of the spacer material. ΔL and ΔT are the change in the length and temperature of the cavity. A comparison of the thermal expansion coefficient of different materials is given in Table 2.1. Ultralow expansion (ULE) glass has a very small thermal expansion coefficient. It also has a zero-crossing (ZC) of its linear thermal expansion coefficient close to room temperature. For ULE-glass with the values given in Table 2.1, the temperature stability of $1\,\text{mK}$ can have relative length stability of the order of 10^{-17}:

$$\frac{\Delta \nu}{\nu} = \frac{\Delta L}{L} = \alpha_{\text{ULE}} \times \Delta T \tag{2.53}$$

One can maintain the cavity at the ZC temperature of the spacer by using Peltiers for active thermal feedback. Also, multiple thermal isolation layers can be used for minimization of the effect of thermal expansion on the spacer to almost a level where the Brownian motion of the coating material dominates instead of the linear thermal expansion of the spacer material.

Table 2.1: Linear expansion coefficient of a few materials. Zero-crossing temperature refers to the temperature value where the linear expansion of the material turns minimum (close to zero), and the thermal expansion coefficient changes sign on either side of this temperature.

	Thermal expansion coefficients	
Material	α-$\times 10^{-6}$ K^{-1} (20° C)	**Zero-crossing temperature**
ULE-glass	$0.0 \pm 0.3 \times 10^{-7}$	5-35° C
Fused-silica glass	0.5	–
Borosilicate glass	3.3	–
Silicon	2.56	4K, 124K
Zerodur	0.007–0.1	5–35° C
BK7 glass	8.5	–
Aluminium	23.1	–

– *Pressure variations:* Pressure changes between the cavity mirrors produce fluctuations in the effective length of the cavity via a change in the refractive index of the cavity medium

$$\frac{\Delta L}{L} = -\left(\frac{dn}{dP}\right)\Delta P. \tag{2.54}$$

For dry air at atmospheric conditions and at room temperature ($\approx$ 300 K), the estimated value is $\frac{dn}{dP} \approx 3 \times 10^{-7}$ mbar^{-1}. The relative frequency stability of the cavity is

$$\frac{\Delta\nu}{\nu} = -(3 \times 10^{-7}\ \text{mbar}^{-1})\Delta P. \tag{2.55}$$

Therefore, for $\frac{\Delta L}{L} < 10^{-14}$ in air requires $\Delta P < 10^{-7}$ mbar. However, attaining such low-pressure fluctuations in atmospheric conditions is challenging. Creating a vacuum environment between the mirrors of the cavity offers a solution to reduce pressure fluctuations. This can be accomplished by using a spacer with a drilled hole, allowing the air between the mirrors to be evacuated within a vacuum system (see Figure 2.7). Thus, the refractive index changes due to air density fluctuations between the mirrors can be reduced by creating an ultrahigh vacuum environment. This is done by isolating the cavity inside a vacuum at the level of 10^{-8} mbar.

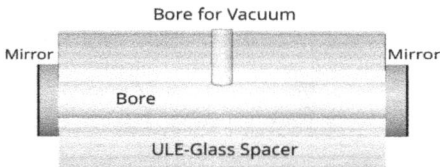

Figure 2.7: ULE-glass spacer used for optical contacting two mirrors for a rigid and stable Fabry–Perot cavity. A bore at the center of the ULE-spacer and at the side is for creating a vacuum between the mirrors once the cavity is placed inside a vacuum chamber.

- *Mechanical vibrations* can severely limit the frequency stability of a cavity. Therefore, achieving ultrastable cavity performance requires passively and actively damping out the vibrations and implementing a particular mounting configuration. The mounting design is such that the cavity spacer is least sensitive to deformation due to external accelerations. Figure 2.8 demonstrates how a cylindrical spacer can have the least deformation if a holding at neutral points is used
  Reducing the influence of external vibration sometimes also requires using air-suspended optical benches. Also, viscoelastic materials (Sorbothone) are used as shock absorbers between the cavity vacuum and the platform holding the vacuum. Active vibration isolation (AVI) platforms can be used to further reduce the effect of external vibration by placing the cavity vacuum on these platforms. An AVI has both passive and feedback strategies to mitigate the impact of vibrations. Passive isolation consists of a series of spring systems. At the same time, active actuators are used for feedback correction against external vibrations. The external vibrations are measured by using acceleration sensors, and the error signal is fed to the actuators.

2.2.2 Tunable cavities and active feedback

To minimize the impact of external perturbations on the cavity length, active feedback can be implemented. Active feedback involves adjusting the cavity length by moving the end mirror using a piezo actuator. By applying a voltage to the piezo actuator, it can

a)

b)

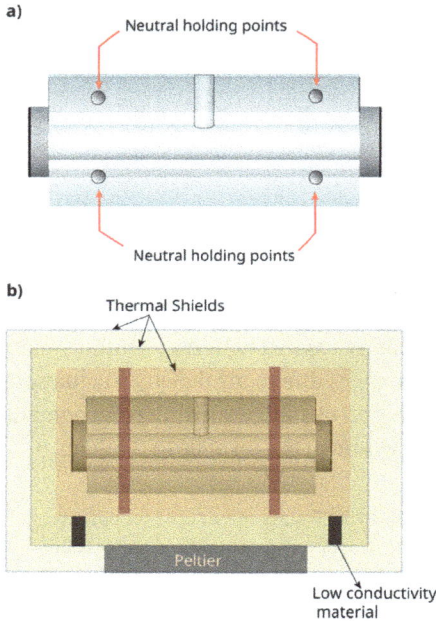

Figure 2.8: Figure (a) shows that the mounting of a cavity can be optimal at specific points called neutral points, for which the length changes of the spacer along the cavity axis under external vibrations are minimal. Figure (b) shows that a cavity is isolated with layers of metallic shields (thermal shields) connected via a very low thermal conductivity material. Along with an active temperature actuator, i. e., peltier, rigorous efforts for thermal isolations are required so that the cavity can be isolated from external temperature changes.

expand or compress, resulting in a proportional displacement of the mirror. In a linear operation regime, the voltage applied to a piezo can produce proportional displacement by expansion or the compression of the piezo material depending on the sign of the applied voltage. However, compensating for the frequency drift of the cavity requires a feedback error signal, which is proportional to the variation of the cavity length over time.

For the active feedback-based cavity, an error signal is necessary to indicate changes in the resonance frequency. This is accomplished by using a stabilized laser and monitoring the cavity's response, either through transmission or reflection on a photodiode. In [12, 13], it has been shown that attaching a piezo to one side of the spacer followed by the end mirror can be used to achieve the frequency stability of the order 10^{-14}. However, further improvements in these kinds of systems are limited by the various noises from the actuator itself.

As explained in the following section, the frequency deviation of the cavity from the laser can be corrected by sending a proportional negative feedback signal to the cavity actuator.

2.2.3 Ultrastable lasers

Active feedback discussed before can be used to either stabilize a laser frequency to an ultrastable cavity resonance or to keep a tunable cavity resonance locked close to a prestabilized laser frequency.

Stabilization of a laser to an ultrastable cavity by active servo feedback is widely used in precision measurement experiments such as atomic clocks. The stabilization technique for this involves transferring the length stability of a cavity to the frequency stability of a laser. This means reading out the frequency drifts of the laser with respect to the cavity resonance and performing fast feedback on the laser using actuators such as the current of the laser diode. In some cases, more than one actuator (piezo and diode temperature) is used for compensating both the fast and slow frequency fluctuations of different frequencies and magnitudes. Extreme care should be taken to isolate any unwanted electronic signal being sent as a feedback correction signal.

Frequency drift estimation requires continuous monitoring of a laser frequency with respect to cavity resonance. From equation (2.8), the cavity transmission is a Lorentzian function and changes symmetrically on both sides as the frequency of the input laser changes. In principle, on the linear part of the slopes of the Lonertzian curve, a change in cavity transmission can be used as an error signal for active feedback on the laser frequency. This technique was used in the initial days for active feedback. However, one of the drawbacks of side-of-fringe locking is that this method cannot differentiate between the intensity changes in the cavity transmission from the laser or the frequency drifts of the cavity. Another method for frequency stabilization of a laser and a cavity is called as the Pound–Drever–Hall locking technique. This technique is one of the most frequently used techniques for error signal generation and active feedback for frequency stabilization.

2.2.4 Pound–Drever–Hall locking method

In optics experiments, it is very often required to have active frequency stabilization, which typically refers to:
- Either keeping the frequency of a laser fixed relative to a stable cavity resonance. In this case, the feedback error signal is applied to the laser actuators, e. g., laser current, temperature, piezo, or any other actuator elements that control the laser's frequency.
- Or for actively correcting a cavity's length (or resonance frequency) drifts relative to a stable laser frequency. Typically, a piezo or any other straining actuator on the cavity can be used for drift corrections.

One of the most frequently used techniques for stabilizing a laser to a cavity (or vice versa) is the Pound–Drever–Hall locking technique [8], which was initially developed for gravitational-wave interferometry [9]. The error signal using this method is produced by creating phase sidebands on the laser (carrier) frequency and their interaction with the cavity. These sidebands are generated either by modulating the laser current or, more conveniently, by using an external electro-optics modulator(EOM), Figure 2.9. The method works as follows.

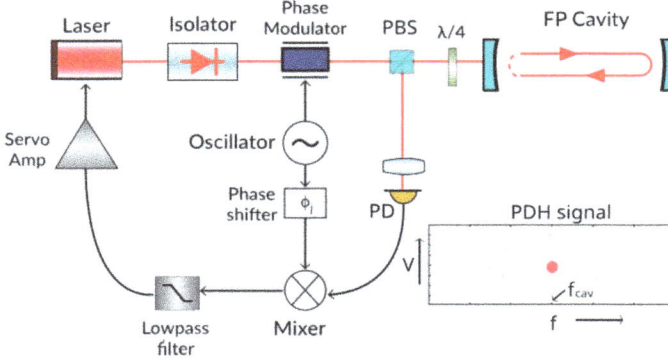

Figure 2.9: Schematics of a Pound–Drever–Hall setup. A laser beam is phase modulated using an EOM, and the light reflected from the cavity is detected on a photodiode and mixed down with EOM drive frequency. The low-pass filtering of the mixed-down signal generates a PDH-error signal.

When the phase modulated laser beam is detected directly on a photodiode, ideally, there is no amplitude modulated signal because the beat of the two sidebands with the carrier cancels each other. This is because of the antisymmetric and equal phase of sidebands with respect to the carrier frequency. However, when this light is incident on a cavity, and the reflected light is detected on a photodiode, an asymmetric phase shift of the sideband produces a beat signal. The phase shift is proportional to the offset from the cavity resonance frequency for small frequency deviations. Suppose the beat signal from the photodiode is mixed down with the RF source (local oscillator) driving the phase modulator, then the resulting signal provides a measure of the frequency difference between the laser and the cavity resonance frequency. The mixed-down signal is also low-pass frequency filtered to remove high-frequency components from the mixing process.

The main advantage of the PDH method is that the generated error signal is the derivative of the cavity transmission spectrum (Lornetzian signal with respect to the detuning $\Delta = \nu_{laser} - \nu_{cavity}$). This allows correct signs of the feedback signal directly on both sides of the cavity resonance. Also, using a reflected signal in the PDH scheme, the servo loop bandwidth is not limited by the cavity response time, which would be the case for a side-of-fringe locking technique using cavity transmission.

Consider a laser beam with angular frequency ω and electric field amplitude E_0. After the phase modulation, the light beam is incident on the cavity. The amplitude of the incident light can be written as follows [5]:

$$E_{in} = E_0 \exp i(\omega t + \beta \sin \Omega t) = (J_0(\beta) + 2iJ_1(\beta) \sin \Omega t) \exp i\omega t$$
$$= E_0(J_0(\beta) \exp i\omega t + J_1(\beta) \exp i(\omega + \Omega)t - J_1(\beta) \exp i(\omega - \Omega)t) \qquad (2.56)$$

To obtain the response of the light reflected from the cavity, we need the reflection transfer function of the cavity. For simplification, we consider $r_1 = r_2 = r$, $\phi = \frac{\omega L}{c} = \frac{\omega}{2\nu_{FSR}}$

in equation (2.15). The resulting transfer function for the cavity reflection is as follows:

$$R_{\text{FP}}(\omega) = \frac{r(\exp(i\frac{\omega}{v_{\text{FSR}}}) - 1)}{1 - r^2 \exp(i\frac{\omega}{v_{\text{FSR}}})} \tag{2.57}$$

The amplitude of the reflection signal from the cavity is

$$E_r = R_{\text{FP}}(\omega)E_{in} \tag{2.58}$$

$$E_r = E_0[R_{\text{FP}}(\omega)J_0(\beta)e^{i\omega t} + R_{\text{FP}}(\omega + \Omega)J_1(\beta)e^{i(\omega+\Omega)t} - R_{\text{FP}}(\omega - \Omega)J_1(\beta)e^{i(\omega-\Omega)t}] \tag{2.59}$$

The reflected power on the photodiode is $|E_r|^2$:

$$
\begin{aligned}
P_r = {} & P_c|R_{\text{FP}}(\omega)|^2 + P_s(|R_{\text{FP}}(\omega + \Omega)|^2 + |R_{\text{FP}}(\omega - \Omega)|^2) \\
& + 2\sqrt{P_c P_s}(\text{Re}(R_{\text{FP}}(\omega)R^*_{\text{FP}}(\omega + \Omega) - R^*_{\text{FP}}(\omega)R_{\text{FP}}(\omega - \Omega)) \cos\Omega t \\
& + (\text{Im}(R_{\text{FP}}(\omega)R^*_{\text{FP}}(\omega + \Omega) - R^*_{\text{FP}}(\omega)R_{\text{FP}}(\omega - \Omega)) \sin\Omega t \\
& + (2\Omega \text{ terms}),
\end{aligned}
\tag{2.60}
$$

where P_c and P_s are the power in the carrier and sideband components respectively.

There are several frequency components. Demodulating the photodiode signal with the local oscillator rf-signal $P_{\text{rf}} \propto \sin\Omega t$ will result in a DC-component and $\cos 2\Omega t$ terms. After low-pass filtering, the error signal from PDH techniques is as follows:

$$\text{err} = 2\sqrt{P_c P_s}\left(\text{Im}(R_{\text{FP}}(\omega)R^*_{\text{FP}}(\omega + \Omega) - R^*_{\text{FP}}(\omega)R_{\text{FP}}(\omega - \Omega))\right) \sin\Omega t \tag{2.61}$$

Figure 2.9 shows the error signal plot with respect to the laser frequency. If the modulating frequency $\Omega \gg \Delta v_c$ where Δv_c is the cavity linewidth, the above equation can be expressed as

$$\text{err} = D\delta\omega, \quad \text{where } D = \frac{8P_c P_s}{\Delta v_c} \tag{2.62}$$

The PDH error signal is linearly proportional to the detuning from the cavity resonance. This error signal is sent to a servo PID loop filter and fed into the actuator for frequency correction. Laser actuators can be a diode current, piezo voltage, or temperature.

2.3 Finesse measurement of a cavity

The finesse of a cavity is an important parameter. It indicates the total number of round trips a photon makes inside the optical resonator before it is lost, either via transmission out of the cavity or due to the absorption or scattering out of the cavity mode. Typically, the finesse ($F = \text{FSR}/\Delta v_{\text{FWHM}}$) estimation involves measuring the cavity linewidth Δv_{FWHM}. Free spectral rang (FSR) is a geometrical parameter that depends on the cavity length.

The cavity linewidth is the full-width at half-maximum (FWHM) linewidth of the Lorentzian spectral lineshape of the resonator ($\Delta\nu_{\mathrm{FWHM}}$). It is directly related to the photon decay time τ_c of the resonator,

$$\tau_c = \frac{1}{2\pi\Delta\nu_{\mathrm{FWHM}}}. \tag{2.63}$$

For linewidth below one MHz, decay time τ_c ($\approx$ few μs to a few hundreds of nanoseconds) can be easily measured with a photodiode signal in the cavity transmission port and a digital oscilloscope. Experimentally, it involves locking a laser frequency at cavity resonance using, e. g., the PDH-locking technique (Section 2.2.4). When the laser beam is abruptly interrupted, the cavity transmission shows an exponential decay of the cavity field leaking through the transmission side. This method is known as the cavity-ring-down scheme. As shown in Figure 2.10, an exponential fit to the decay curve of the cavity transmission reveals the linewidth of the cavity.

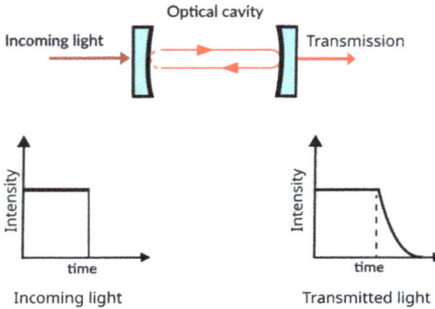

Figure 2.10: The figure shows the cavity-ring down method for measuring the cavity decay time for linewidth estimation. By abruptly switching off the input light, the cavity transmission signal shows an exponential decay of the cavity field, which directly relates to the cavity linewidth.

It is interesting to note that the microresonators have cavity linewidth in the range of a few tens of MHz to a few GHz due to their small size. Therefore, a more suitable method for cavity linewidth estimation in those cases involves using EOM generated sideband as a frequency marker on the injected laser beam and then estimating the width of the cavity resonance.

2.4 Gravitational wave-detection and cavity interferometers

Gravitational waves are produced by gigantic cosmic objects having nonaxis symmetric rotation (e. g., a merger of black holes [2]) [1, 3, 4]. They have been successfully detected via optical interferometric techniques and are one of the most remarkable and finest examples of extreme precision in experimental physics. Detection of gravitational waves requires measurements of the length strain of the order of $\frac{\Delta L}{L} \sim 10^{-21}$. To intuitively comprehend this measurement, one can imagine measuring a displacement with 10^{-18} for 1 km of length or $\frac{1}{1000}$ times the size of a proton. For such high-precision measurement, the signal generated from the stretching and contracting of the interferometer's two arms has to be integrated appropriately so as not to wash out the signal due to too much integration or not to miss out on the signal itself due to too little integration. Using optical resonators and interferometric techniques, length strain of the order of 10^{-21}, has been detected with a certainty of 99.9994 %. Therefore, a very high level of technical challenges has been overcome to achieve the sensitivity of the interferometer for detecting the GW waves. Here, we present a short description of the experimental schematics to show the extent and the importance of the resonator techniques used in various sections and stages of the experimental setup. Here, in no way do we attempt to describe the exquisite and involved experimental system but rather to point out the role of various optical resonators used in enhancing the detection capability of the GW detectors.

Figure 2.11 shows a schematic layout of the various optical cavities used in the LIGO detector. In brief, a laser beam is first frequency stabilized and mode cleaned using a tri-

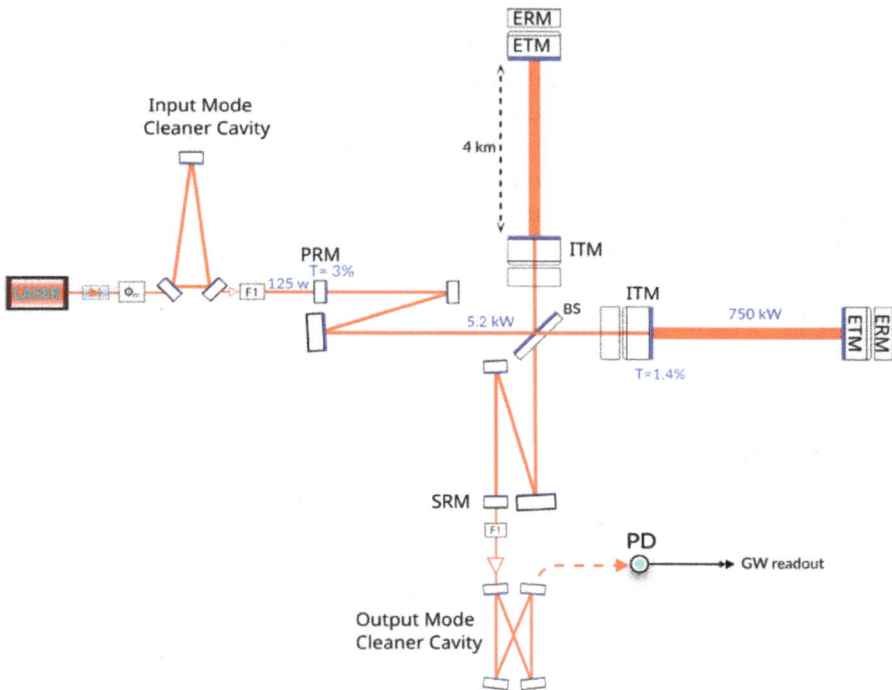

Figure 2.11: A schematic depiction of the laser interferometer-based gravitational wave detector. The abbreviations used for various mirrors are PRM = power recycling mirror, ERM = end recycling mirror, ETM = end test mass, ITM = input test mass, and SRM = signal recycling mirror. PD is the photodiode. In the experiment, a laser beam is frequency stabilized and mode cleaned using input-mode-cleaner cavity. The beam is injected in the interferometer using PRM and the power gain is obtained in the two arms of the interferometer using arm cavities. Finally, the output signal is detected on a PD after a SRM and an output-mode-cleaner cavity. Most of the optical components are operated under vacuum environment. This image is adapted from [6] © IOP Publishing. Reproduced with permission. All rights reserved.

angular cavity at the input side of the interferometer. The beam then enters a Michelson interferometer via the beam splitter (BS) after transmitting through a power-recycling mirror (PRM). Using end-recycling mirrors (ERM) and ITM, a power gain by a factor of a few hundred is achieved along the interferometer arms. Finally, the interferometer output is set to the dark fringe configuration, and any differential drift in the lengths of the two arms is detected on PD after the leaking light is transmitted through a signal-recycling mirror (SRM) and an output-mode-cleaner cavity. In the following, a short description of some of these cavities is given to understand multiple possible applications of optical cavities in precision measurements.

LIGO detector

LIGO is an acronym for Large Interferometric Gravitational Wave Antenna. LIGO consists of an ultrahigh precision Michelson interferometer capable of measuring fractional length changes between the two arms with precision for strain measurement $\approx 10^{-21}$. As the gravitational waves produce a transversal shear strain, an ultrahigh precision Michelson interferometer is a natural detector choice to measure such strains. The main concept behind this is that a Michelson interferometer can measure the differential displacement of its two end mirrors with extremely high sensitivity. This high level of sensitivity is possible by reducing the shot-noise level in the detection process and improving the SNR of the GW signal using various techniques in optical and electronic signal detection and processing. Optical resonators, one of the crucial optical devices that has driven much of the development in laser science and various other domains of experimental physics, are critical in enhancing GW-signal detection sensitivity at different stages of the LIGO interferometer.

2.4.1 Arm cavity

The optical cavity and Michelson interferometer both work on the principle of the interference of light waves. An optical cavity is a high-sensitivity phase detector, while a Michelson interferometer can be used to have common mode rejection and high-sensitivity differential length measurement between the two arms by using laser light. By combining both technologies, one can have higher phase sensitivity and an effectively increased optical path length for longer interaction of light with the Gravitational strain.

Arm cavities are used in each of the arms of the interferometer to increase the effective length of the Michelson interferometer in the LIGO detector (Figure 2.12). The cavity used in each arm of the interferometer thus effectively enhances the phase shift on the laser light produced by a change in the arm lengths of the interferometer as the GW-wave passes. Each cavity is a Fabry–Perot cavity consisting of an input mirror (input test mass, ITM) and an end mirror (end test mass, ETM). The end mirrors are designed

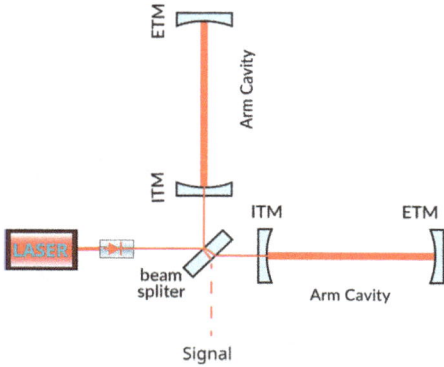

Figure 2.12: Simplified schematics of the Fabry–Perot cavities used along the interferometer arms to increase the effective length of the interferometer and to obtain power gain in the GW-detector.

to have the highest reflectivity possible. This minimizes the transmission losses in the cavity via end mirrors. The input mirrors parameters are chosen such that each arm cavity is highly overcoupled. Therefore, the reflected signal from the arm cavity is dominantly the field leaking out of the cavity and not the light directly reflected from the input mirror without entering the cavity. Arm cavities have a finesse of a few hundred. Intensity and frequency noise, which occurs on a time scale faster than the cavity storage time, are filtered away. Therefore, an important parameter is the pole frequency of the resonator. The pole frequency is determined by the storage time within the cavity and depends on the finesse of the cavity:

$$f_{\text{pole}} = \frac{1}{2\pi\tau_s}, \quad \tau_s = \left(\frac{2L}{\pi c}\right)F, \tag{2.64}$$

where L is the length of the resonator, c is the speed of light, and F is the finesse of the resonator.

2.4.2 Power recycling cavity

The LIGO detector measures the strain produced by the gravitational waves by measuring the relative path-length change between the two arms of a highly specialized Michelson interferometer. Typically, the detection is done at the dark fringe configuration at the beam splitter of the interferometer. It means that close to the operational point, most of the laser light is reflected back toward the laser after returning from the interferometer arms.

The power recycling cavity is used to recycle the back reflected light from the interferometer. This is done by placing a partially reflective mirror between the laser and the beam splitter of the interferometer and recycling the laser right back into the interferometer (Figure 2.11). Based on the total round-trip loss at the mirror, one can choose the reflectivity of the recycling mirror to have maximum light reflected in the interferometer and minimum leakage light toward the laser. Therefore, power built-up using a recycling cavity is achieved by maintaining appropriate resonance conditions in the interferometer and the recycling cavity. Considering the Michelson interferometer as a compound mirror with complex reflectivity of r_{MI} and recycling mirror reflectivity as r_{RM} and transmissivity t, then the effective power gain from the recycling cavity can be given as

$$G = \left(\frac{t}{1 - r_{RM} r_{MI}} \right)^2 \tag{2.65}$$

For low-loss recycling mirrors, the gain of the recycling cavity is typically limited by the losses of the Michelson interferometer.

By using a finesse of a few tens for the recycling cavity, one can obtain circulating optical power close to hundreds of watts with an input laser power of a few tens of watts.

2.4.3 Premode cleaner cavity

The laser beam in LIGO detector is transmitted through a mode-cleaner cavity before being injected into the interferometer. The primary function of this cavity is to filter the spatial mode of the light coming out of the laser. This cavity also passively filters out the laser's power and frequency noise above the cavity pole frequency (equation (2.64)). Therefore, the mode-cleaner cavity is used as a reference cavity to stabilize the frequency and the beam pointing of the laser on the lowest fundamental TEM_{00} mode. By using Pound–Drever–Hall (PDH) locking technique (see Section 2.2.4), the laser is locked to the fundamental cavity resonance. As the laser frequency follows the length stability of the PMC cavity, a great deal of effort goes into the passive and active isolation of the cavity from the various vibration noise sources. In an LIGO detector, a 4-mirror bow-tie-shaped configuration is used with an FSR of 150 MHz and finesse of around 130. For the high stability of the cavity, it is isolated inside a vacuum with reduced external vibration sensitivity. One of the advantages of using a triangular or bow-tie-shaped cavity is that the main reflection from the cavity can be directed away from the incoming beam direction. Therefore, the choice of angle of incidence on the end mirror is a trade-off between reducing the backscattered light and the mode ellipticity arising due to the increased angle of incidence. Also, the acquired Gouy phase of the PMC cavity allows the frequency of the fundamental TEM_{00} mode to be well separated from the higher-order modes such that they are mostly reflected from the cavity.

2.4.4 Output-mode cleaner cavity

Although the spatial mode of the laser beam is cleaned before injecting it into the interferometer using a premode cleaner cavity, the beam transversal through various optical components of the interferometer and their deformations can cause higher or non-fundamental modes to arise. The beam passes through the output-mode cleaner cavity (OMC) to avoid these higher-order modes reaching the photodiode at the antisymmetric port of the interferometer (Figure (2.11)). OMC, therefore, improves the shot noise by reducing the detection noises, such as higher-order spatial modes and side-band fields. There are different cavity configurations proposed for this purpose. One of the cavity systems is a four-mirror bow-tie cavity arrangement. The folded configuration allows having a relatively compact design due to folded beam path. Also, the transmission through the two mirrors (apart from the input and piezo-mounted mirrors) can be used for diagnostic purposes. To have the optimum transmission of the interferometer signal through OMC, it should be stabilized at the frequency of the interferometer output beam. Passive stabilization of the cavity is obtained by optimal vibration isolation inside the vacuum, while the active stabilization of the cavity length enables to achieve a significant reduction in the shot-noise of the GW-detector.

2.4.5 Signal extraction cavity

In some of the configurations for the gravitational wave detectors, signal recycling is implemented by adding a mirror (signal mirror) at the output, i. e., the dark port of the interferometer. By using a signal recycling mirror after the interferometer and before the photodiode at the dark port of the interferometer (see Figure 2.11), the GW signal can be extracted more efficiently from the interferometer [16, 11]. The cavity formed by this mirror, as shown in Figure 2.13, is used to have a broader frequency response of the interferometer. This can be understood based on the following arguments.

When the output port of the interferometer is configured to have destructive interference (dark port), any light signal results at this dark port only when there is a differential change in arm lengths. Therefore, the signal mirror sees the light only for

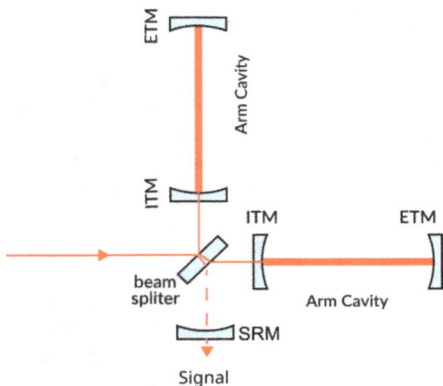

Figure 2.13: By placing a signal recycling mirror (SRM) at the output port of a Michelson interferometer, one can have a detection bandwidth tunability as described in the main text.

differential changes in the interferometer arms. The output signal is consequently integrated by the signal mirror and strongly depends on the optical properties of the signal mirror. Effectively, the signal mirror allows for the optimization of various gravitational wave signals of interest. Additionally, the resulting signal cavity also can help reduce the amount of stray light leaking out of the dark port.

One of the other advantages of using a signal mirror in the interferometer is more laser light storage in the arm cavities. This is because, in the absence of the signal mirror, one can optimize the frequency response of the interferometer by only adjusting the transmittance of the arm-cavity end mirrors. Therefore, in the absence of a signal recycling mirror, a limitation arises on the end mirror reflectivity and the power stored in the arm cavities due to the constrained mirror properties. On the other hand, with a signal mirror in place, the signal bandwidth can be much broader, and one can choose the arm-cavity end mirrors independently to have more power gain. Therefore, adding a signal mirror allows more freedom in selecting the interferometer optics.

Bibliography

[1] B. P. Abbott, R. Abbott, T. D. Abbott, M. R. Abernathy, F. Acernese, K. Ackley, C. Adams, T. Adams, P. Addesso, R. X. Adhikari, and et al. Gw151226: observation of gravitational waves from a 22-solar-mass binary black hole coalescence. *Physical Review Letters*, 116(24):241103, 2016.

[2] B. P. Abbott, R. Abbott, T. D. Abbott, M. R. Abernathy, F. Acernese, K. Ackley, C. Adams, T. Adams, P. Addesso, R. X. Adhikari, and et al. Observation of gravitational waves from a binary black hole merger. *Physical Review Letters*, 116(6):061102, 2016.

[3] B. P. Abbott, R. Abbott, T. Abbott, F. Acernese, K. Ackley, C. Adams, T. Adams, P. Addesso, R. X. Adhikari, V. B. Adya, and et al. Gw170817: observation of gravitational waves from a binary neutron star inspiral. *Physical Review Letters*, 119(16):161101, 2017.

[4] R. Abbott, T. D. Abbott, S. Abraham, F. Acernese, K. Ackley, C. Adams, R. X. Adhikari, V. B. Adya, C. Affeldt, M. Agathos, and et al. Gw190521: A binary black hole merger with a total mass of $150 M_\odot$. *Physical Review Letters*, 125(10):101102, 2020.

[5] E. D. Black. An introduction to pound–drever–hall laser frequency stabilization. *American Journal of Physics*, 69(1):79–87, 2001.

[6] The LIGO Scientific Collaboration, J. Aasi, B. P. Abbott, R. Abbott, T. Abbott, M. R. Abernathy, K. Ackley, C. Adams, T. Adams, P. Addesso, R. X. Adhikari, V. Adya, C. Affeldt, N. Aggarwal, O. D. Aguiar, A. Ain, P. Ajith, A. Alemic, B. Allen, D. Amariutei, S. B. Anderson, W. G. Anderson, K. Arai, M. C. Araya, C. Arceneaux, J. S. Areeda, G. Ashton, S. Ast, S. M. Aston, P. Aufmuth, C. Aulbert, B. E. Aylott, S. Babak, P. T. Baker, S. W. Ballmer, J. C. Barayoga, M. Barbet, S. Barclay, B. C. Barish, D. Barker, B. Barr, L. Barsotti, J. Bartlett, M. A. Barton, I. Bartos, R. Bassiri, J. C. Batch, C. Baune, B. Behnke, A. S. Bell, C. Bell, M. Benacquista, J. Bergman, G. Bergmann, C. P. L. Berry, J. Betzwieser, S. Bhagwat, R. Bhandare, I. A. Bilenko, G. Billingsley, J. Birch, S. Biscans, C. Biwer, J. K. Blackburn, L. Blackburn, C. D. Blair, D. Blair, O. Bock, T. P. Bodiya, P. Bojtos, C. Bond, R. Bork, M. Born, S. Bose, P. R. Brady, V. B. Braginsky, J. E. Brau, D. O. Bridges, M. Brinkmann, A. F. Brooks, D. A. Brown, D. D. Brown, N. M. Brown, S. Buchman, A. Buikema, A. Buonanno, L. Cadonati, J. Calderón Bustillo, J. B. Camp, K. C. Cannon, J. Cao, C. D. Capano, S. Caride, S. Caudill, M. Cavaglià, C. Cepeda, R. Chakraborty, T. Chalermsongsak, S. J. Chamberlin, S. Chao, P. Charlton, Y. Chen, H. S. Cho, M. Cho, J. H. Chow, N. Christensen, Q. Chu, S. Chung, G. Ciani, F. Clara, J. A. Clark, C. Collette, L. Cominsky, M. Constancio, D. Cook, T. R. Corbitt, N. Cornish, A. Corsi, C. A. Costa, M. W. Coughlin, S. Countryman, P. Couvares, D. M. Coward,

M. J. Cowart, D. C. Coyne, R. Coyne, K. Craig, J. D. E. Creighton, T. D. Creighton, J. Cripe, S. G. Crowder, A. Cumming, L. Cunningham, C. Cutler, K. Dahl, T. Dal Canton, M. Damjanic, S. L. Danilishin, K. Danzmann, L. Dartez, I. Dave, H. Daveloza, G. S. Davies, E. J. Daw, D. DeBra, W Del Pozzo, T. Denker, T. Dent, V. Dergachev, R. T. DeRosa, R. DeSalvo, S. Dhurandhar, M. D'ıaz, I. Di Palma, G. Dojcinoski, E. Dominguez, F. Donovan, K. L. Dooley, S. Doravari, R. Douglas, T. P. Downes, J. C. Driggers, Z. Du, S. Dwyer, T. Eberle, T. Edo, M. Edwards, M. Edwards, A. Effler, H. -B. Eggenstein, P. Ehrens, J. Eichholz, S. S. Eikenberry, R. Essick, T. Etzel, M. Evans, T. Evans, M. Factourovich, S. Fairhurst, X. Fan, Q. Fang, B. Farr, W. M. Farr, M. Favata, M. Fays, H. Fehrmann, M. M. Fejer, D. Feldbaum, E. C. Ferreira, R. P. Fisher, Z. Frei, A. Freise, R. Frey, T. T. Fricke, P. Fritschel, V. V. Frolov, S. Fuentes-Tapia, P. Fulda, M. Fyffe, J. R. Gair, S. Gaonkar, N. Gehrels, L. Á. Gergely', J. A. Giaime, K. D. Giardina, J. Gleason, E. Goetz, R. Goetz, L. Gondan, G. González, N. Gordon, M. L. Gorodetsky, S. Gossan, S. Go3ler, C. Gräf, P. B. Graff, A. Grant, S. Gras, C. Gray, R. J. S. Greenhalgh, A. M. Gretarsson, H. Grote, S. Grunewald, C. J. Guido, X. Guo, K. Gushwa, E. K. Gustafson, R. Gustafson, J. Hacker, E. D. Hall, G. Hammond, M. Hanke, J. Hanks, C. Hanna, M. D. Hannam, J. Hanson, T. Hardwick, G. M. Harry, I. W. Harry, M. Hart, M. T. Hartman, C-J. Haster, K. Haughian, S. Hee, M. Heintze, G. Heinzel, M. Hendry, I. S. Heng, A. W. Heptonstall, M. Heurs, M. Hewitson, S. Hild, D. Hoak, K. A. Hodge, S. E. Hollitt, K. Holt, P. Hopkins, D. J. Hosken, J. Hough, E. Houston, E. J. Howell, Y. M. Hu, E. Huerta, B. Hughey, S. Husa, S. H. Huttner, M. Huynh, T. Huynh-Dinh, A. Idrisy, N. Indik, D. R. Ingram, R. Inta, G. Islas, J. C. Isler, T. Isogai, B. R. Iyer, K. Izumi, M. Jacobson, H. Jang, S. Jawahar, Y. Ji, F. Jiménez-Forteza, W. W. Johnson, D. I. Jones, R. Jones, L. Ju, K. Haris, V. Kalogera, S. Kandhasamy, G. Kang, J. B. Kanner, E. Katsavounidis, W. Katzman, H. Kaufer, S. Kaufer, T. Kaur, K. Kawabe, F. Kawazoe, G. M. Keiser, D. Keitel, D. B. Kelley, W. Kells, D. G. Keppel, J. S. Key, A. Khalaidovski, F. Y. Khalili, E. A. Khazanov, C. Kim, K. Kim, N. G. Kim, N. Kim, Y. -M. Kim, E. J. King, P. J. King, D. L. Kinzel, J. S. Kissel, S. Klimenko, J. Kline, S. Koehlenbeck, K. Kokeyama, V. Kondrashov, M. Korobko, W. Z. Korth, D. B. Kozak, V. Kringel, B. Krishnan, C. Krueger, G. Kuehn, A. Kumar, P. Kumar, L. Kuo, M. Landry, B. Lantz, S. Larson, P. D. Lasky, A. Lazzarini, C. Lazzaro, J. Le, P. Leaci, S. Leavey, E. O. Lebigot, C. H. Lee, H. K. Lee, H. M. Lee, J. R. Leong, Y. Levin, B. Levine, J. Lewis, T. G. F. Li, K. Libbrecht, A. Libson, A. C. Lin, T. B. Littenberg, N. A. Lockerbie, V. Lockett, J. Logue, A. L. Lombardi, L. Lormand, J. Lough, M. J. Lubinski, H. Lück, A. P. Lundgren, R. Lynch, Y. Ma, J. Macarthur, T. MacDonald, B. Machenschalk, M. MacInnis, D. M. Macleod, F. Magaña-Sandoval, R. Magee, M. Mageswaran, C. Maglione, K. Mailand, I. Mandel, V. Mandic, V. Mangano, G. L. Mansell, S. Márka, Z. Márka, A. Markosyan, E. Maros, I. W. Martin, R. M. Martin, D. Martynov, J. N. Marx, K. Mason, T. J. Massinger, F. Matichard, L. Matone, N. Mavalvala, N. Mazumder, G. Mazzolo, R. McCarthy, D. E. McClelland, S. McCormick, S. C. McGuire, G. McIntyre, J. McIver, K. McLin, S. McWilliams, G. D. Meadors, M. Meinders, A. Melatos, G. Mendell, R. A. Mercer, S. Meshkov, C. Messenger, P. M. Meyers, H. Miao, H. Middleton, E. E. Mikhailov, A. Miller, J. Miller, M. Millhouse, J. Ming, S. Mirshekari, C. Mishra, S. Mitra, V. P. Mitrofanov, G. Mitselmakher, R. Mittleman, B. Moe, S. D. Mohanty, S. R. P. Mohapatra, B. Moore, D. Moraru, G. Moreno, S. R. Morriss, K. Mossavi, C. M. Mow-Lowry, C. L. Mueller, G. Mueller, S. Mukherjee, A. Mullavey, J. Munch, D. Murphy, P. G. Murray, A. Mytidis, T. Nash, R. K. Nayak, V. Necula, K. Nedkova, G. Newton, T. Nguyen, A. B. Nielsen, S. Nissanke, A. H. Nitz, D. Nolting, M. E. N. Normandin, L. K. Nuttall, E. Ochsner, J. O'Dell, E. Oelker, G. H. Ogin, J. J. Oh, S. H. Oh, F. Ohme, P. Oppermann, R. Oram, B. O'Reilly, W. Ortega, R. O'Shaughnessy, C. Osthelder, C. D. Ott, D. J. Ottaway, R. S. Ottens, H. Overmier, B. J. Owen, C. Padilla, A. Pai, S. Pai, O. Palashov, A. Pal-Singh, H. Pan, C. Pankow, F. Pannarale, B. C. Pant, M. A. Papa, H. Paris, Z. Patrick, M. Pedraza, L. Pekowsky, A. Pele, S. Penn, A. Perreca, M. Phelps, V. Pierro, I. M. Pinto, M. Pitkin, J. Poeld, A. Post, A. Poteomkin, J. Powell, J. Prasad, V. Predoi, S. Premachandra, T. Prestegard, L. R. Price, M. Principe, S. Privitera, R. Prix, L. Prokhorov, O. Puncken, M. Pürrer, J. Qin, V. Quetschke, E. Quintero, G. Quiroga, R. Quitzow-James, F. J. Raab, D. S. Rabeling, H. Radkins, P. Raffai, S. Raja, G. Rajalakshmi, M. Rakhmanov, K. Ramirez, V. Raymond, C. M. Reed, S. Reid, D. H. Reitze, O. Reula, K. Riles, N. A. Robertson, R. Robie, J. G. Rollins, V. Roma, J. D. Romano, G. Romanov, J. H. Romie, S. Rowan, A. Rüdiger, K. Ryan, S. Sachdev, T. Sadecki, L. Sadeghian, M. Saleem, F. Salemi, L. Sammut,

V. Sandberg, J. R. Sanders, V. Sannibale, I. Santiago-Prieto, B. S. Sathyaprakash, P. R. Saulson, R. Savage, A. Sawadsky, J. Scheuer, R. Schilling, P. Schmidt, R. Schnabel, R. M. S. Schofield, E. Schreiber, D. Schuette, B. F. Schutz, J. Scott, S. M. Scott, D. Sellers, A. S. Sengupta, A. Sergeev, G. Serna, A. Sevigny, D. A. Shaddock, M. S. Shahriar, M. Shaltev, Z. Shao, B. Shapiro, P. Shawhan, D. H. Shoemaker, T. L. Sidery, X. Siemens, D. Sigg, A. D. Silva, D. Simakov, A. Singer, L. Singer, R. Singh, A. M. Sintes, B. J. J. Slagmolen, J. R. Smith, M. R. Smith, R. J. E. Smith, N. D. Smith-Lefebvre, E. J. Son, B. Sorazu, T. Souradeep, A. Staley, J. Stebbins, M. Steinke, J. Steinlechner, S. Steinlechner, D. Steinmeyer, B. C. Stephens, S. Steplewski, S. Stevenson, R. Stone, K. A. Strain, S. Strigin, R. Sturani, A. L. Stuver, T. Z. Summerscales, P. J. Sutton, M. Szczepanczyk, G. Szeifert, D. Talukder, D. B. Tanner, M. Tápai, S. P. Tarabrin, A. Taracchini, R. Taylor, G. Tellez, T. Theeg, M. P. Thirugnanasambandam, M. Thomas, P. Thomas, K. A. Thorne, K. S. Thorne, E. Thrane, V. Tiwari, C. Tomlinson, C. V. Torres, C. I. Torrie, G. Traylor, M. Tse, D. Tshilumba, D. Ugolini, C. S. Unnikrishnan, A. L. Urban, S. A. Usman, H. Vahlbruch, G. Vajente, G. Valdes, M. Vallisneri, A. A. van Veggel, S. Vass, R. Vaulin, A. Vecchio, J. Veitch, P. J. Veitch, K. Venkateswara, R. Vincent-Finley, S. Vitale, T. Vo, C. Vorvick, W. D. Vousden, S. P. Vyatchanin, A. R. Wade, L. Wade, M. Wade, M. Walker, L. Wallace, S. Walsh, H. Wang, M. Wang, X. Wang, R. L. Ward, J. Warner, M. Was, B. Weaver, M. Weinert, A. J. Weinstein, R. Weiss, T. Welborn, L. Wen, P. Wessels, T. Westphal, K. Wette, J. T. Whelan, S. E. Whitcomb, D. J. White, B. F. Whiting, C. Wilkinson, L. Williams, R. Williams, A. R. Williamson, J. L. Willis, B. Willke, M. Wimmer, W. Winkler, C. C. Wipf, H. Wittel, G. Woan, J. Worden, S. Xie, J. Yablon, I. Yakushin, W. Yam, H. Yamamoto, C. C. Yancey, Q. Yang, M. Zanolin, F. Zhang, L. Zhang, M. Zhang, Y. Zhang, C. Zhao, M. Zhou, X. J. Zhu, M. E. Zucker, S. Zuraw and J. Zweizig. Advanced ligo. *Classical and Quantum Gravity*, 32(7):074001, Mar 2015. https://doi.org/10.1088/0264-9381/32/7/074001.

[7] J. Davila-Rodriguez, F. N. Baynes, A. D. Ludlow, T. M. Fortier, H. Leopardi, S. A. Diddams, and F. Quinlan. Compact, thermal-noise-limited reference cavity for ultra-low-noise microwave generation. *Optics Letters*, 42(7):1277–1280, Apr 2017. https://doi.org/10.1364/OL.42.001277. URL https://opg.optica.org/ol/abstract.cfm?URI=ol-42-7-1277.

[8] R. W. P. Drever, J. L. Hall, F. V. Kowalski, J. Hough, G. M. Ford, A. J. Munley, and H. Ward. Laser phase and frequency stabilization using an optical resonator. *Applied Physics B*, 31:97–105, 1983.

[9] R. W. P. Drever, G. M. Ford, J. Hough I. M. Kerr, A. J. Munley, J. R. Pugh, N. A. Robertson, and H. Ward. A gravity-wave detector using optical cavity sensing. *General relativity and gravitation 1980*, page 265, 1983.

[10] A. D. Ludlow, X. Huang, M. Notcutt, T. Zanon-Willette, S. M. Foreman, M. M. Boyd, S. Blatt, and J. Ye. Compact, thermal-noise-limited optical cavity for diode laser stabilization at 1×10−15. *Optics Letters*, 32(6):641–643, Mar 2007. https://doi.org/10.1364/OL.32.000641. URL https://opg.optica.org/ol/abstract.cfm?URI=ol-32-6-641.

[11] J. E. Mason and P. A. Willems. Signal extraction and optical design for an advanced gravitational-wave interferometer. *Applied Optics*, 42(7):1269–1282, 2003.

[12] K. Möhle. Piezoelectrically tunable optical cavities for the gravitational wave detector lisa, 2013.

[13] K. Möhle, E. V. Kovalchuk, K. Döringshoff, M. Nagel, and A. Peters. Highly stable piezoelectrically tunable optical cavities. *Applied Physics B*, 111(2):223–231, May 2013. ISSN 1432-0649. https://doi.org/10.1007/s00340-012-5322-0.

[14] J. M. Robinson, E. Oelker, W. R. Milner, W. Zhang, T. Legero, D. G. Matei, F. Riehle, U. Sterr, and J. Ye. Crystalline optical cavity at 4 K with thermal-noise-limited instability and ultralow drift. *Optica*, 6(2):240–243, Feb 2019. https://doi.org/10.1364/OPTICA.6.000240. URL https://opg.optica.org/optica/abstract.cfm?URI=optica-6-2-240.

[15] A. E. Siegman. *Lasers*. University science books, 1986.

[16] N. D. Smith. Techniques for improving the readout sensitivity of gravitational wave antennae. Ph. D. Thesis, 2012.

[17] S. A. Webster, M. Oxborrow, S. Pugla, J. Millo, and P. Gill. Thermal-noise-limited optical cavity. *Physical Review A*, 77:033847, Mar 2008. https://doi.org/10.1103/PhysRevA.77.033847. URL https://link.aps.org/doi/10.1103/PhysRevA.77.033847.

3 Fiber-based resonators

The previous two chapters introduce two key concepts: guided propagation of the electromagnetic wave in optical fibers and the technology of optical resonators (cavities). While optical fibers are the essential building blocks for light transport in optical experiments and the telecommunication industry, optical resonators provide the means to confine and control light waves. Combining these technologies open up possibilities for a fiber-based system that integrates small resonators. This integration allows for the direct incorporation of cavity-enabled optical devices and sensors in a compact platform compatible with fiber technology and suitable for integration into fiber networks.

In this chapter, we focus on fiber-based resonator systems, exploring their design and characterization process. One question arises: what qualifies as a fiber-based resonator system? In principle, any microresonator with input and output ports directly connected to the fibers can be considered as such. However, for our purposes, we will only include microresonators that are created by directly modifying the fibers themselves in this category (see Section 3.2).

3.1 Why fiber resonators

Optical resonators play a vital role in various fields of experimental physics and optics. In the previous chapter, we explored the remarkable stability and finesse exhibited by macroscopic bulk optical resonators, which have been utilized in high-precision experiments such as ultrastable lasers and gravitational wave detectors. Conversely, microresonators offer the advantage of incredibly small mode volumes, often approaching the wavelength of light λ, along with high finesse. This makes them suitable for interfacing with atoms and molecules in quantum technology. The tight focus of light and the high-quality factor of these microresonators enable strong interaction with the medium inside, even reaching the quantum realm of single-photon and single-atom interactions.

Therefore, microresonators are a critical area of research. Their capability of strong confinement and enhancement of light within a well-defined mode structure enables the development of compact-size and low-power photonic devices and circuits. By changing the shape, size, and material properties, microresonators can be produced for specific applications of optical devices. For example, microcavities with extremely high-quality factors and very small mode volumes are particularly researched for their application in quantum technology and optoelectronics devices. The small mode volume and high finesse of these lead to enhanced light-matter interaction, improved detection sensitivity, and nonlinear phenomenon due to power enhancement.

Fiber-based microresonators are particularly intriguing due to their inherent compatibility with fiber-coupled input–output channels and their alignment-free handling of high-quality resonators. These systems have been extensively explored in the context

https://doi.org/10.1515/9783110636260-003

of quantum technology and quantum networks. Thanks to their small size and compatibility with fiber networks, they offer the possibility of integrating quantum sensors and interfaces seamlessly into fiber-based quantum networks using photons.

3.2 Different fiber-based microresonators

Fiber-based optical resonators encompass various types that are created by directly modifying standard optical fibers. In this section, we will introduce three common types: Fiber Fabry–Perot cavity, nanofiber resonators, and whispering gallery mode resonators. We will provide a brief overview of each resonator system, and subsequently delve into the fabrication and characterization processes in the following sections.

–   Fiber Fabry–Perot cavities (FFPC) consist of two directly fabricated mirrors on the fiber-end facets. To form a stable cavity, at least one of the fiber mirrors is concave and curved. The small size of fibers allow easy optical [25] access from the side in comparison to a macroscopic cavity with similar performance (see Figure 3.1). This accessibility is crucial when introducing matter inside the cavity and requiring optical interaction.

–   Micro(nano)-fiber resonators are produced by elongation and tapering of a standard fiber to a very small diameter (see Figure 3.2). Typically, the fiber diameter can range from close to a micrometer to a few hundred nanometers. The resonator is formed by having Bragg mirrors at the ends of the waist on both sides [29]. These resonators have a strong evanescent field of light near the surface of the waist of the

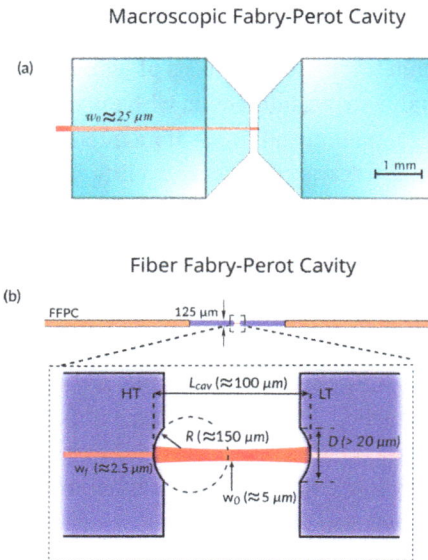

Figure 3.1: Figure (a) shows a high finesse macroscopic Fabry–Perot cavity with optimized size for a minimum cavity beam waist of 25 µm. In contrast to this, Figure (b) demonstrates the dimensions of a high finesse fiber Fabry–Perot-cavity producing a beam waist of 5 µm.

Fiber Fabry-Perot Cavity   Nano-Fiber Cavity   Fiber Bottle Cavity

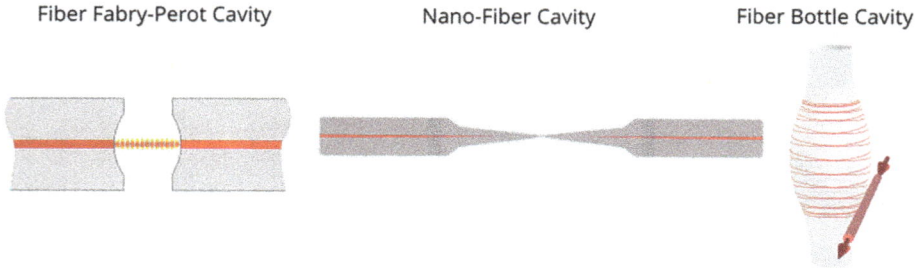

Figure 3.2: This figure demonstrates three distinct optical resonators created by directly modifying the conventional single-mode fibers.

fiber. Strong light-matter interaction can be realized near the waist and this forms the basis of strong light-matter interaction via evanescent light coupling.

– Fiber-based Whispering Gallery Mode (WGM) resonators are constructed by tapering a regular fiber in a specific pattern, e. g., to have a bottle-shaped WGM resonators [18] (see Figure 3.2) where light is confined because of the total internal reflection within this structure.

3.3 Fabrication of the fiber Fabry–Perot cavity

To achieve a Fabry–Perot cavity with high-quality factor, mirrors with extremely high reflectivity are required. For instance, some high-performance cavities have mirrors with reflectivity $R \approx 0.999998$. Such high reflectivity can only be accomplished with the mirror substrate having surface roughness close to the atomic scales and coatings with losses within a few parts per million (ppm). Thus, the initial step in constructing a high finesse fiber Fabry–Perot cavity (FFPC) involves creating a curved mirror surface on a fiber tip. The surface roughness must be low enough to have a negligible contribution to mirror losses and coating reflectivity. Creating these miniaturized mirrors on fiber end facets, with diameters of approximately ~125 μm, presents a challenging task. One of the first research works on this was presented in [26], where the research group demonstrated the fabrication of curved mirror surfaces using the following approach. First, they generated smooth microlens on a planar silica substrate. Subsequently, they affixed a fiber tip to the surface of the microlens. Finally, a lift-off technique involving the application of force to the fiber allowed for the transfer of the curved mirror to the fiber tip, as depicted in Figure 3.3. However, the finesse reported was only ~ 1000.

Later, the much more efficient process of making micromirror surfaces using CO_2 laser machining directly on a fiber tip was demonstrated. The resulting cavity had a finesse ~ 130000. Subsequently, FFPCs consisting of laser-machined mirrors gained popularity after the demonstration of the strong interaction of light with matter inside the cavity [25, 3]. FFPCs have already been used for coupling various types of single quan-

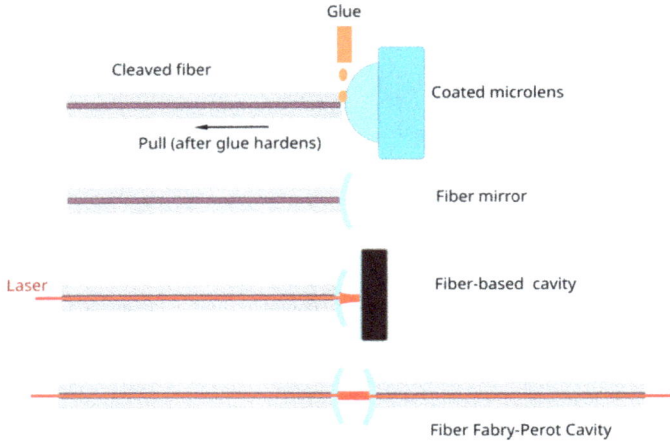

Figure 3.3: One of the early demonstrations of fiber-based cavities involved gluing fiber ends to curved glass surfaces and lifting off a smooth curved surface on the fiber tips [26]. The fiber tips are then coated with dielectric layers to form high-reflectivity mirrors.

tum emitters to the cavity mode. This has paved the way for exciting applications of fiber cavity-based quantum electrodynamics (QED) experiments for miniaturized quantum devices, e. g., single-photon sources, quantum memories, and quantum dynamical systems (see Chapter 4). In the following sections, we will discuss the design, fabrication, assembly, and characterization process of the fiber Fabry–Perot cavity.

3.3.1 Laser machining of glass

A CO_2 laser is a versatile tool used in various industrial applications, including glass processing. When it comes to generating very smooth surfaces on glass, a CO_2 laser can be employed through a process known as laser polishing (or machining). Laser machining with a CO_2 laser involves selectively heating the surface of the glass to a high temperature, causing localized melting and reflow of the material. The intense energy from the laser beam is absorbed by the glass, leading to rapid heating and subsequent material transformation. The molten glass surface then undergoes controlled cooling, resulting in the desired smoothness and improved surface quality.

As the fabrication of very smooth microcurved surfaces on a fiber tip is required to create a high-quality mirror for FFPCs, this can be created by laser machining and polishing techniques for glass surfaces using a CO_2 laser. Typically, an optical experimental setup with a CO_2 laser is used to generate spatial and temporal modulated laser pulses on a fiber surface. Because of the strong absorption of CO_2 laser on glass, curved surfaces with different depths and radii of curvature (tens of micrometers to a few millimeters) can be created, either by melting or by material removal (see the section below). It is

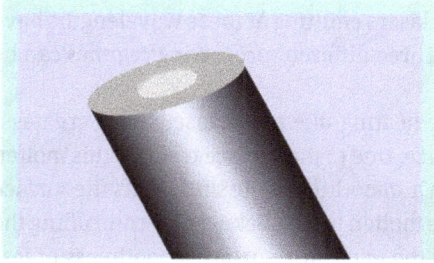

Figure 3.4: Artistic impression of a single-mode fiber tip after laser machining. The curved depression at the center of the fiber tip is machined by glass ablation technique using a CO_2 laser.

possible to achieve surface roughness below 1 nm (see Figure 3.4) with relative ease and requires control over the laser power and pulse width. No direct contact with the substrate surface is required compared to mechanical polishing methods, which makes this method perfectly suitable for creating curved surfaces even on fibers having diameters of hundred micrometers. Figure 3.5 shows the absorption curve for the silica glass at infrared laser radiation [11]. It can be seen that the absorption is very strong near the

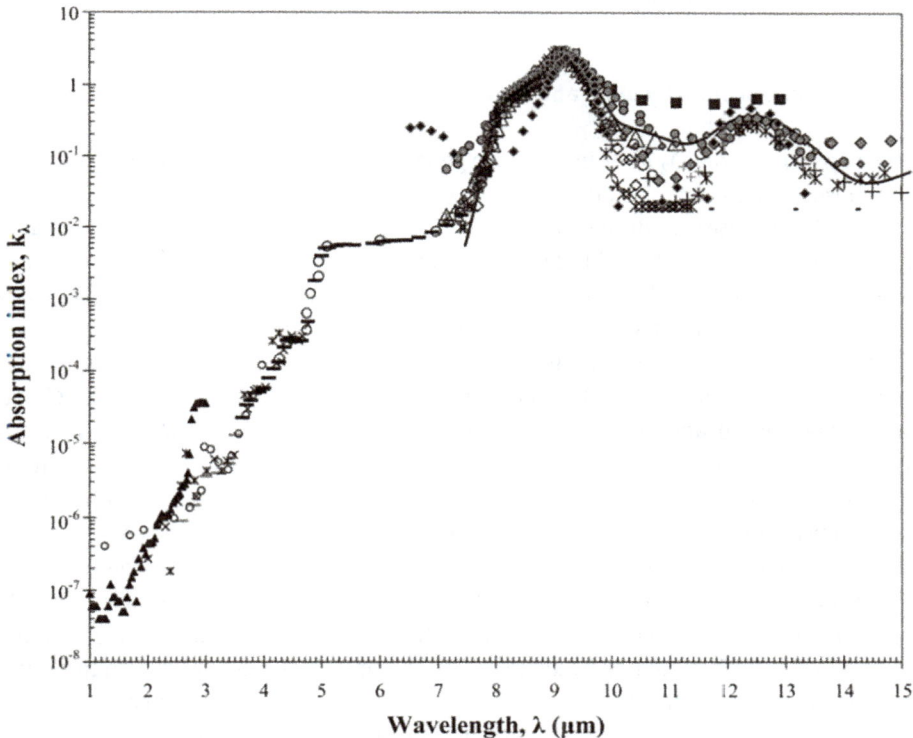

Figure 3.5: Graph shows the approximate absorption profile of glass for laser wavelengths between 1–15 μm. The absorption peaks around 9.3 μm wavelength of the CO_2 laser. Reprinted from [11] with permission © The Optical Society.

9–10.6 µm wavelength regime. Therefore, CO_2 lasers emitting at these wavelengths have been used for the surface treatment of glass. Three different processing regimes can be identified:

– At low powers (few hundred milliwatts), heating due to the laser beam creates a very thin molten layer of glass at the surface. Due to the surface tension, this molten layer spreads over the surface resulting in a smoothing of the surface. As the surface smoothening is roughly comparable to the molten layer thickness, by controlling the incident laser power and the duration of the exposure, surface smoothness of less than 1 nm is possible.

– By using high-intensity laser pulses, a large domain of the glass, much deeper inside the surface of the substrate, can be melted. In this regime, one can create a convex structure on a glass substrate. The above is due to the surface minimizing effect from the surface tension of the melted glass. A thin molten layer also settles over the curved surface producing a surface smoothening effect.

– For high-intensity and short laser pulses, it is possible to remove glass material by evaporation, resulting in a deep concave structure with ultralow smoothness. This process is known as glass ablation and is discussed in the next section.

3.3.2 Microfabrication by glass ablation

Strong absorption of CO_2 laser in glass can generate very high temperatures for a suitably chosen pulse energy and duration. For example, in a particular regime of parameters, the laser pulses result in material removal due to the direct evaporation of glass. This process, known as glass ablation, is a robust technique to create very smooth concave surfaces on glass substrates and is particularly suitable for machining fiber tips. The very smooth surfaces with surface roughness below 1 nm can be created with robustness against small fluctuations in laser parameters. Therefore, micromirrors on fiber tips can be created easily by placing fibers on a laser beam and then controlling the laser power, duration of the pulses, and the size of the laser beam waist incident on the fiber tip. The laser power, pulse duration, and waist determine the curvature's depth and radius. Typically, laser pulses are used, which have high enough intensity and power to generate high temperature (3000 K) at the glass surface. The vapor pressure of glass at such a high temperature exceeds the ambient pressure and leads to high evaporation rates. Also, surface melting allows a smooth surface on an atomic scale. A correct adjustment of laser pulse parameters leads to a thin molten glass layer being flown over the surface created by ablation. For example, if one uses very high intensity and very short pulses, the melted glass layer solidifies before it can spread over the surface. On the other hand, the melted glass layers can extend too far into the volume for very long pulses. This can create an unwanted convex shape toward the edges of the concave surface. In [25], laser pulses with a pulse ranging from a few ms to 100 ms, with an average power of 0.2–2 watts, have been used to create smooth curved surfaces with a radius of curvature varying from 20 µm–500 µm, which uses 10.6 µm CO_2 laser. For the Fabry–Perot cavities with very small mode volume, glass ablation is one of the easier options. The traditional super polishing techniques are difficult to apply for very small-size mirrors required for cavities with small-mode volumes. An approximate analysis of the glass ablation process and the achievable parameters for the concave surface can be performed by considering the surface absorption of the laser pulse. Considering the Gaussian intensity distribution for the laser pulses,

$$I = I_0 \exp\left(\frac{-2r^2}{w^2}\right)$$ (3.1)

where I_0 is the peak intensity and w is the beam waist of the laser. If the glass has an absorption coefficient of A, thermal conductivity κ, and a thermal diffusion coefficient of D, then the resulting temperature profile on the glass surface is calculated using a two-dimensional heat flow equation. The temperature profile on the glass surface, for a laser pulse of duration τ is as follows [5]:

$$T(r, \tau) = \frac{Aw^2 I_0}{2\sqrt{\pi}} \frac{\sqrt{D}}{\kappa} \left(\int_a^b \frac{e^{-\left(\frac{r^2}{\frac{w^2}{2} + 4Dy}\right)}}{\sqrt{y}(\frac{w^2}{2} + 4Dy)} \, dy \right)$$ (3.2)

The maximum temperature on the glass surface, i.e., at the center of the incident beam is as follows:

$$T(0, \tau) = \frac{Aw^2 I_0}{2\sqrt{\pi}} \arctan \sqrt{\frac{8D\tau}{w^2}}.$$ (3.3)

For pulses of duration, τ, the depth of the structure can be expressed as follows:

$$z(r) = z_0 \exp\left(\frac{-U}{k_b T(0, r)} \frac{r^2}{2w^2}\right).$$ (3.4)

The resulting structure on the glass is approximately a Gaussian surface. The central region of the curved surface can be approximated by a spherical depression and the diameter of which is given by [5]:

$$d = 2w_0 \sqrt{\frac{k_b T(0, \tau)}{U}}.$$ (3.5)

Although this expression assumes an infinite plane for heat diffusion and ignores any nonlinearities and finite-size effects, it correctly approximates the created concave surface on extended substrates. However, for the concave surface on the fiber tip deviates from this simple model due to the finite size effect in transverse heat conduction.

3.3.3 Fiber mirror production setup

The following steps are typically followed to create micromirror structures on the fiber end facets:

– The process starts with creating a clean and optically flat fiber end-facet. This is obtained by the precision cleaving of a fiber which involves making an orthogonal cut and applying tensile force along the fiber symmetry axis, (see Figure 3.6). The resulting breaking of the fiber with the cleave process creates a very smooth optical surface at the fiber tip. Commercial cleavers are available and provide high repeatability and a high-quality optical surface. The quality of the cleave is determined by the obtained surface uniformity over most of the fiber-end-facet surface and the angle of the surface with respect to the fiber axis. Commercial cleavers can be used to create surface flatness of very high optical quality. The cleaved surfaces with angles as small as 0.3 deg are possible by commercially available cleavers [14]. These small angles are necessary to avoid a decentration between the center of the mirror and the fiber core.

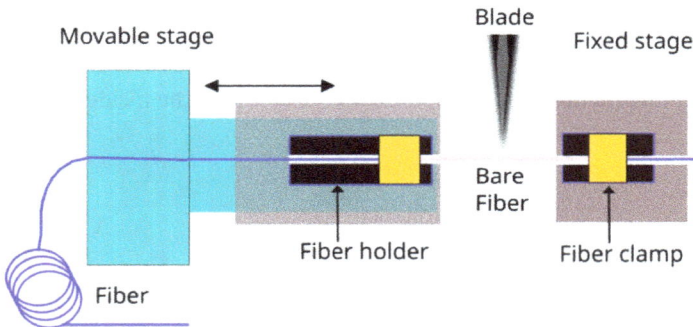

Figure 3.6: Schematics of a fiber cleaving process.

As a standard, optical fiber is coated with a protective layer of polymer or metal coating to reduce external stress, removing the protective coating is important before cleaving the fiber. This involves removing the protective layer either by mechanical stripping using commercial fiber stripper or by a chemical dissolving processes. The fiber is thereafter cleaned using Isopropanol. The stripped bare fiber is now placed between the two clamping points of the cleaver, and a low tensile force is applied. A fast moving diamond blade is then used to cut the fiber precisely while pulling the fiber along its axis. If proper tension is applied, the desired cleave surfaces are high-quality optical surfaces with very small angles with respect to the fiber's symmetry axis. Improper tension and bad impact with the diamond blade can produce non-uniform and angled surfaces. Angled surfaces are avoided because the glass ablation process creates a mirror curvature that is off-center with respect to the fiber core. The above reduces the mode-matching between the fiber core and the cavity mode.

– Cleaved fiber is then mounted on a holder and can be inspected for clean cleaving and flatness using an interference microscope. The interferometric techniques used for analyzing the fiber surfaces before and after laser processing are described in Section 3.4. The difference between the interferograms, for two different regimes of cleaving, are shown in Figure 3.7. The left picture shows a badly cleaved fiber with a significant cleave angle and surface distortion producing multiple fringe patterns, while the right image shows a uniform, flat cleaved surface with almost a single bright fringe at the center of the fiber surface.

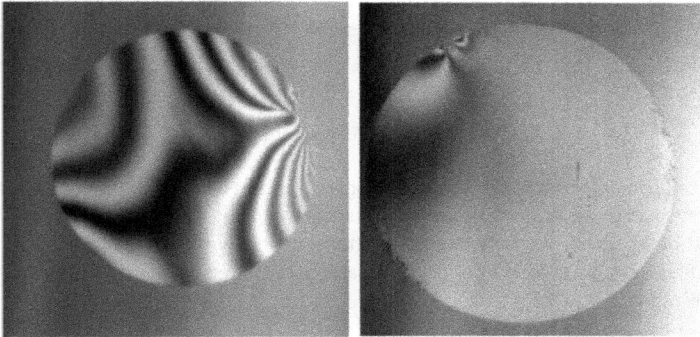

Figure 3.7: Interference image for two different surface qualities of a cleaved fiber end facet. A badly cleaved flat surface (left) with distorted and multiple fringe pattern, and a good cleaved fiber surface with almost a single uniform bright fringe (right).

– Cleaved fibers are then required to be brought to the laser machining area, where appropriately timed laser pulses are used. The laser ablation process, as described above, creates a very smooth concave mirror template. The laser setup used for the laser machining is described as follows.

Laser machining of fiber tips

Figure 3.8 shows the experimental setup to create mirror templates on a fiber-end facet. A mode-cleaned CO_2 laser beam is focused and incident on the glass substrate after spatial filtering using a pinhole. A mechanical shutter or an Acousto-optics-modulator (AOM) is used for generating pulses of the laser. Metal-coated mirrors and high-power compatible Zinc-sellnide optics (lenses) are used for focusing and deflecting the laser on the substrate. A beam waist of a few tens to a few hundred micrometers is used depending on the desired mirror curvature. A small part of the laser can be split and detected

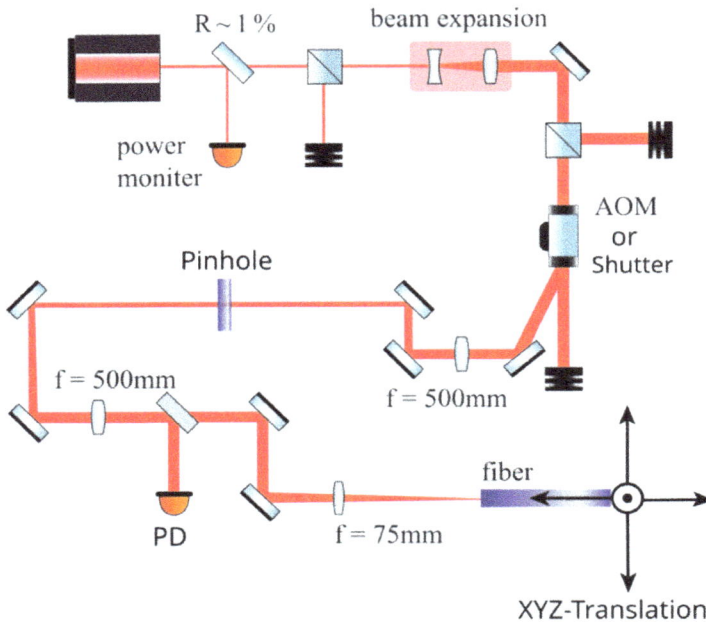

Figure 3.8: Experimental setup for laser machining fiber tips using a CO_2 laser. Following the laser, beam-shaping optics are utilized to manipulate the laser beam's characteristics. Additionally, a device for pulse generation is incorporated into the setup, such as a mechanical shutter or an acoustic-optic modulator (AOM). AOM is driven by rf-power and is used for controlling the laser pulse power and duration by controlling the rf-power fed to the AOM.

on a photodetector for either laser power monitoring or active intensity stabilization using AOM and a servo. For precise fiber placement with respect to the laser beam's waist, a high-precision three-dimensional translation stage is used. The travel precision on the translation stage, below 1 μm, is used with a requirement of high positional reproducibility. Also, larger travel distances are also required to move the target between the inspection region and the laser shooting (machining region). A computer interface can automate the process of moving fiber between different regions, shining temporal laser pulses, and analyzing the surface.

The different research groups reported laser power between 1–10 watts, with a waist diameter between 10–200 μm, and 1–100 pulses with a temporal pulse duration of 1–100 ms [25]. The experimental parameters for laser machining can be optimized for each setup based on the target parameters for the mirror surfaces. After each laser processing of the fiber surfaces, one requires high precision analysis of the surfaces. In the experiment, this analysis is done by moving the substrate to the inspection microscope. The inspection microscope is an interferometric device used for surface analysis and is described in the next section.

3.4 Interferometric techniques for surface metrology

Optical interferometric techniques are extensively utilized for precise surface metrology. In a fiber mirror production process, either a Michelson or a variant, which is a white light interferometer, is used for estimating the surface quality. These techniques provide a fast surface quality evaluation with high reproducibility and repeatability, which are crucial in determining the quality of the surface after laser machining.

The fundamental principle behind interferometric techniques is to compare a target surface with a known reference surface. By analyzing the interference fringe pattern created by the overlapping light reflected from the target surface and the reference surface, the quality of the target surface can be determined. To achieve this, either the object or the reference surface is moved, allowing for surface evaluation at different object depths, in a direction perpendicular to the probing light beam. A brief introduction to these surface analysis techniques is given below.

3.4.1 Michelson interferometer

The schematic of a Michelson interferometer is shown in Figure 3.9 A monochromatic light is split using a beam splitter (BS) into two paths. Both light beams from path A and path B are reflected back to the BS and overlap at the output port C of the beam splitter. The interference of the electric field amplitudes produces a spatial modulation

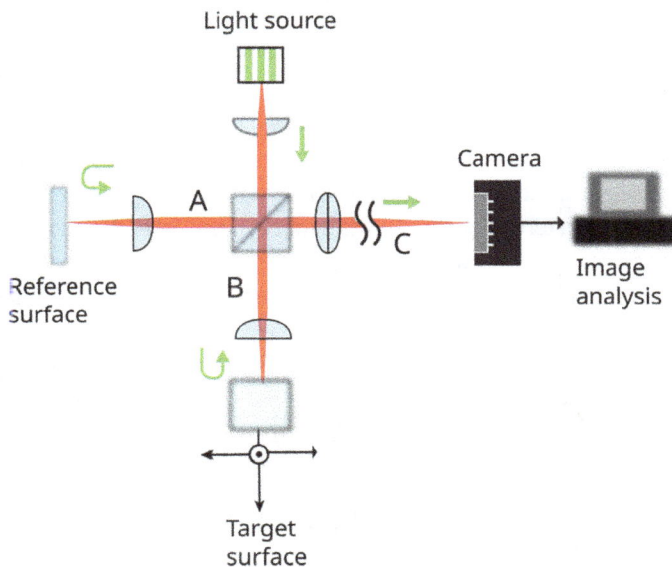

Figure 3.9: Schematic of a Michelson interferometer used for surface analysis of a f ber-end facet. The light reflected from the fiber-end and a reference mirror, interferes on a camera.

of the intensity pattern, which depends on the phase difference arising from the relative difference between the two paths:

$$E_1 = E_{01} \cos(k.x - \omega t)$$
$$E_2 = E_{02} \cos(k.x - \omega t + \phi)$$

The resulting intensity (fringe) pattern for a monochromatic light is

$$I = |E|^2 = I_0 + I_m \cos \phi$$

where I_0 and I_m are the backgrounds and the modulation intensity while the phase ϕ arises due to the difference in the optical path length between the two arms. The surface height variations are directly related the phase ϕ as follows:

$$\Delta h(x,y) = \frac{\lambda}{4\pi} \phi(x,y) \tag{3.6}$$

where λ is the wavelength of the probing light. The data processing techniques that can retrieve the interferogram phase values can be used to calculate the surface profile or deformations.

Suppose a high surface quality reference mirror is used in path A, and a substrate with surface irregularities is placed on path B. In that case, the resulting interference pattern shows a relative deviation of surface B with respect to surface A, Figure 3.10. The intensity variations in the fringe pattern can be mapped to the relative depth variations using the phase-shifting interferometry analysis technique.

3.4.2 Phase-shifting interferometry

The phase-shifting interferometry technique reveals the surface topography by analyzing phase changes in a series of interferograms. As described above, these interferograms are obtained due to the interference between the light reflected from the object's surface and a reference surface during each point of a scan perpendicular to the analysis surface. These fringe patterns are recorded on a CCD camera. Intensity variation on each pixel of the camera is plotted against the scan distances. These scans, as shown in Figure 3.10, are oscillating functions representing the sinusoidal brightness variation along the scan. The phase acquired at a point $((x,y,z)$ is:

$$I(x,y,z) = I_A(x,y) + I_B(x,y) + 2\sqrt{(I_A(x,y)I_B(x,y) \cos(\phi(x,y) + \phi(z)))} \tag{3.7}$$

The phase shift at each location in the x-y plane is related to the depth variation with the following expression:

$$\phi(x,y) = \frac{4\pi\Delta h(x,y)}{\lambda}$$

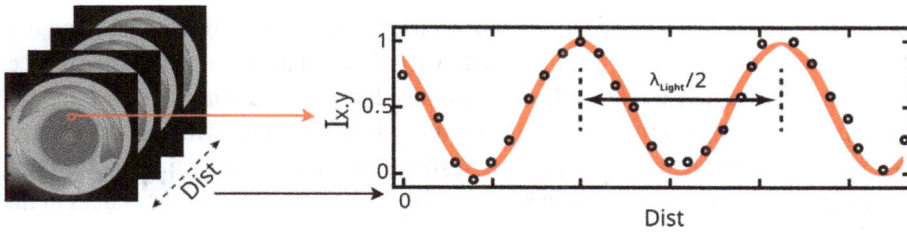

Figure 3.10: Schematics of the steps involved in a phase-shift interferometric technique. A series of inter-ferograms are obtained for the object's surface while scanning either the reference mirror or the object. Pixel-by-pixel reading out of the intensity pattern with respect to the scan provides depth information on the object plane. A sinusoidal fit to the data is performed, and the phase information is extracted from this. An example of an intensity variation of a single pixel (denoted by a red circle) is shown on the right-side figure as a sinusoidal curve.

However, calculating the phases directly from the interferogram results in phase values between $-\pi$ and π. The actual phases directly related to the surface profile are calculated using a phase unwrapping technique [6].

The phase-shifting technique is minimally affected by inhomogenous illumination of the target surface and, therefore, presents an enormous improvement over the direct analysis of a single interference pattern between the light reflected from the reference and the target surface. The limitation of this method arises when adjacent points have a phase difference larger than 2π and, therefore, actual phase difference cannot be ex-tracted unambiguously. This situation arises for a rough surface with a sharp-step struc-ture. For this kind of scenario, phase-shifting interferometry using monochromatic light fails and requires other techniques.

Figure 3.10 shows the brightness of a single pixel for different positions of the refer-ence plate. The axial scan is realized by a high-precision piezo stage. The obtained phase map containing the phase difference can be analyzed by a phase-unwrapping algorithm. The phase map retrieved from this process is shown in Figure 3.10b. The vertical resolu-tion in such a system, in principle, depends on the precision of the translation stage and the sensitivity of the camera. However, in reality, the noise induced by the vibrations in the system limits the measurement precision. Also, discontinuities arising at phase change of 0 and 2π create some artifacts, which can be up to the level of a few tens of nanometers.

3.4.3 White light interferometry

White light interferometry (WLI), also known as vertical scanning white light interfer-ometry, is an interferometric technique used for precise surface topography analysis. It overcomes the phase ambiguity problem of phase-shifting interferometry and achieves high repeatability, providing accurate height measurements down to a few nanometers.

This technique takes advantage of the short coherence of a white light source and employs fringe localization to determine height variations relative to a reference surface. The fringe localization works by finding the position of the maximum of a Gaussian envelope of the fringe pattern and indicates the height variations with respect to a reference surface (see Figure 3.11). One can construct the surface topography by scanning vertically to the object's surface and simultaneously acquiring a series of images of the interference fringe pattern on each pixel of a camera image.

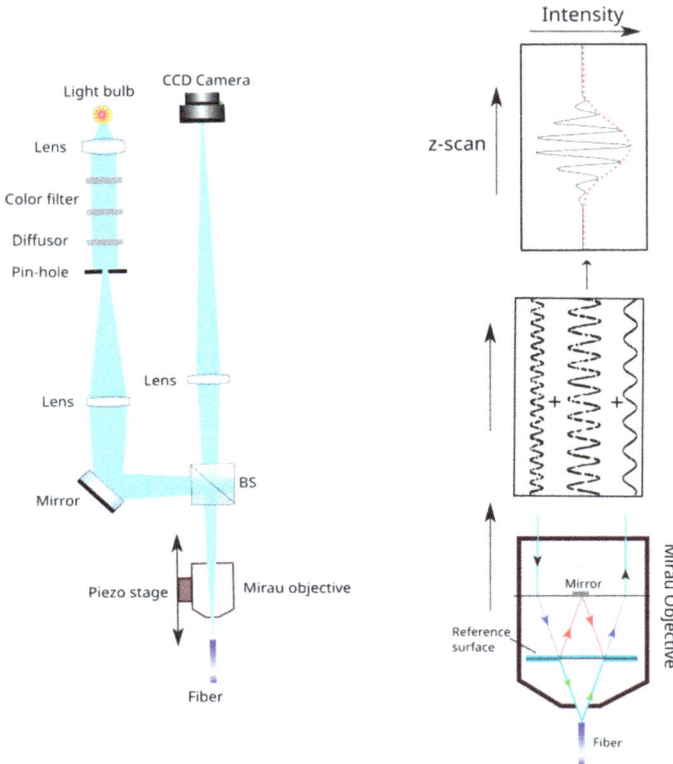

Figure 3.11: Schematics of a white light interferometer setup. The Mirau objective is used to create the surface profile of an object using interferometric techniques.

The experimental setup for white light interferometry (Figure 3.11) typically involves passing the white light through frequency filters to shape its spectrum, creating an approximate Gaussian spectrum within the range detected by a camera (usually around $\lambda \sim$ 400–800 nm). A high-resolution microscope objective is used to generate the interference pattern. Different microscope objectives such as Mirau, Michelson, and Linnik can be employed, each incorporating a partially reflecting reference surface.

The interference pattern between the reflected light from the object's surface and the one from the internal reference surface interferes and forms a fringe pattern for each vertical position of the objective. The high-precision piezo-translation stage performs vertical scanning of the objective. The positional accuracy is typically a few nanometers (closed loop) and the scan step size of tens of nanometers. A fast digital camera captures a series of fringe patterns while the objective scans vertically across the object's surface. These images are then processed and analyzed on a computer to construct the surface topography.

The short coherence length of the light source in WLI ensures that interference fringes are only obtained when the observation beam and the partially reflected beam from an internal mirror have nearly identical optical paths, typically within micrometers, and depend on the coherence of the white light source.

Figure 3.11 (right-image) shows the recorded interferogram, which shows a maximum modulation on the pixel when the path length difference between the two reflected beams is equal. Therefore, by constructing the brightest pixel point for each camera pixel at each scan position, one could reconstruct the structure of the surface along the vertical axis. If the surface is flat, then moving the object will create the brightest pixel at a fixed distance for all camera pixels. For a curved surface, the distance between the maximum modulation point will directly depend on the curvature of the surface. Therefore, white light interferometry can profile surfaces by localization of the fringe created from a low-coherence light source.

During the scanning process, the interference pattern recorded on each pixel of the camera provides valuable information for extracting the height profile of the surface. The fringe contrast at each pixel is sensitive to the difference in optical path length, serving as a reliable measure of the object height relative to the reference surface. The intensity profile of a pixel follows a sinusoidal function modulated by a visibility function. The interference fringes exhibit a Gaussian coherence envelope, characterized by fast-frequency sinusoidal oscillations.

Analytically, the intensity profile on a pixel is a sinusoidal function modulated by a visibility function

$$I(z) = I_0 V(z) \cos \phi. \tag{3.8}$$

For a low-coherence white light source having Gaussian spectral density, with coherence length $l_c = \frac{c}{\pi \Delta v}$, the central wavelength λ_0 and the bandwidth of the light source Δv; the intensity profile is (see [15]):

$$I(z) = I_0 \left[\left(1 + \exp\left(-4 \left(\frac{z - z_0}{l_c} \right)^2 \right) \right) \cos \left(4\pi \frac{z - z_0}{\lambda_0} - \phi_0 \right) \right] \tag{3.9}$$

where z is the vertical position of the object, and z_0 is the corresponding position of the reference surface. From the equation above, it is clear that the coherence envelope function during a vertical scan is a Gaussian function containing fast-frequency sinusoidal oscillations.

Fourier transform can be used to filter out the high-frequency part from the interferogram to obtain the coherence envelope. A conceptual procedure to obtain the coherence envelope is depicted in Figure 3.12. The Fourier transform of the interferogram exhibits three frequency components, the carrier frequency, and the two sideband frequency components, a negative and a positive, with respect to the carrier frequency. Filtering to keep only the positive frequency component allows for extracting the slow-frequency envelope. A Gaussian fit to the envelope function can extract the maximum of the envelope function. More advanced surface reconstruction techniques are used for steeped surfaces and to avoid the ghost pixel point on the surface reconstruction algorithms and can be found in [7, 12].

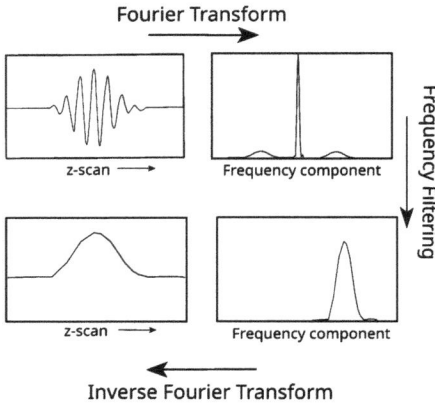

Figure 3.12: A procedural diagram of the Fourier analysis method. The top left figure is a single-pixel interferogram and its Fourier transform is on the top right. The figure on the bottom right is the frequency component, remaining after frequency filtering the carrier and the negative frequency component. The bottom left is the desired envelope function, which is obtained by the inverse Fourier transform of the frequency-filtered FFT. The maximum of the Gaussian envelope function is the location of the brightest pixel on the z-scan positions.

With the continuous progress in fast imaging, acquisition, and computing techniques, WLI offers a fast and reliable surface analysis method. The short coherence time of the broadband white light source used in WLI has many advantages. For example, with the broad spectrum and the short coherence time of the light source, noise due to the spurious interference between the optical components can be reduced as compared to when a laser or a narrow bandwidth filtered light is used. This is because the short coherence allows discernible interference images only when the path lengths are within a few micrometers or less.

3.4.4 Evaluation of mirror surfaces

Once a curved surface is fabricated on the fiber ends through laser machining, it is important to evaluate its quality to determine the mirror properties of the template. This evaluation involves assessing the radius of curvature, mirror diameter, and any ellipticity present along the orthogonal axes of the mirror surface.

After reconstructing the curved surface profile using interferometric imaging, it becomes possible to estimate the radius of curvature and mirror diameter. A two-dimensional Gaussian fit can be applied to the surface, allowing for the assessment of ellipticity along the two perpendicular axes. If the lengths of these axes are different, it indicates the presence of ellipticity in the mirror surface. Additionally, fitting a spherical surface close to the bottom of the curve enables the estimation of the radius of curvature. These evaluations are crucial in determining the mirror properties of the template created through the laser ablation technique. The mirror surfaces produced with low ellipticity and high smoothness are ideal for creating high-quality mirrors intended for high-finesse cavities.

Figure 3.13 illustrates two scenarios where mirror templates exhibit different levels of ellipticity after laser machining. Based on the estimations of geometrical parameters, the laser machining parameters can be adjusted to achieve the desired curved surface. Advanced laser machining techniques, such as multishot techniques (discussed in Section 3.8.1), often involve continuous adjustment of the laser pulse followed by surface evaluation until the desired shape of the curve is achieved.

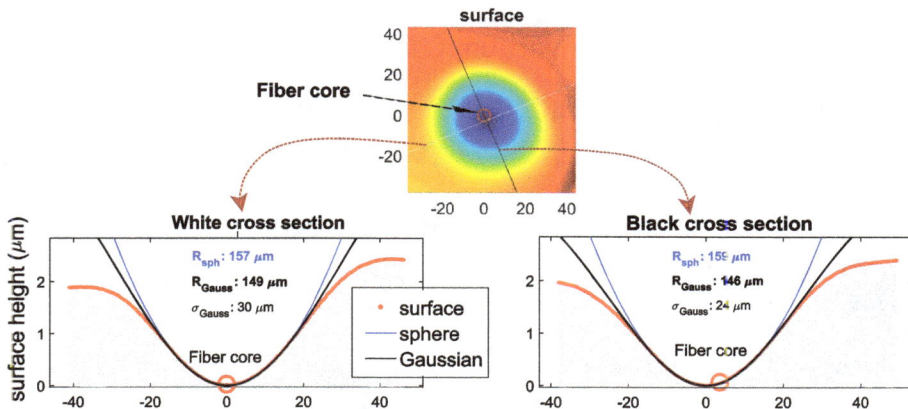

Figure 3.13: The cross-section of a mirror surface along two orthogonal axes is fitted with a spherical and Gaussian function to estimate the radius of curvature of the mirror produced by glass ablation.

3.5 Mirror coating techniques

Mirror coating techniques involve the deposition of precise thin film layers on a substrate to achieve high reflectivity. Among these techniques, the ion-beam sputtering (IBS) method stands out as a typical approach that enables the creation of coatings with minimal losses. The excellent control over coating parameters allows the IBS method to attain losses as low as a few parts per million (typically, $L \sim 1 - 5\,\text{ppm}$), and the lowest achievable transmission $T = 1 - R\,(\sim 5)\,\text{ppm}$. Therefore, IBS method is suitable for high reflectivity dielectric mirror coatings on laser-machined fiber-end facets. Low coating losses ensure that only a minimal fraction of the incident light is absorbed or scattered within the coating. In the IBS method (illustrated in Figure 3.14), a vapor of ionized dielectric material, such as Ta_2O_5 (with a refractive index of 2.28) or SiO_2 (with a refractive index of 1.45) is deposited in alternating layers on the fiber-end facet. The thickness of each layer is chosen to be approximately a quarter of the wavelength of the desired high reflectivity light. The configuration of the coating involves a stack of layers with alternating high and low refractive index coatings. This arrangement is carefully selected to ensure that the reflections from the different layers interfere constructively, resulting in maximum interference for a wide range of incident beam angles. These dielectric-coated mirrors are sometimes referred to as Bragg mirrors due to their ability to exploit the Bragg interference phenomenon.

In the ion-beam sputtering method, precise control over the thickness of the coating layers is achievable. This is made possible by utilizing an ion beam that allows for the accurate ejection of the sputtered material at a level of a few angstroms. The sputtered

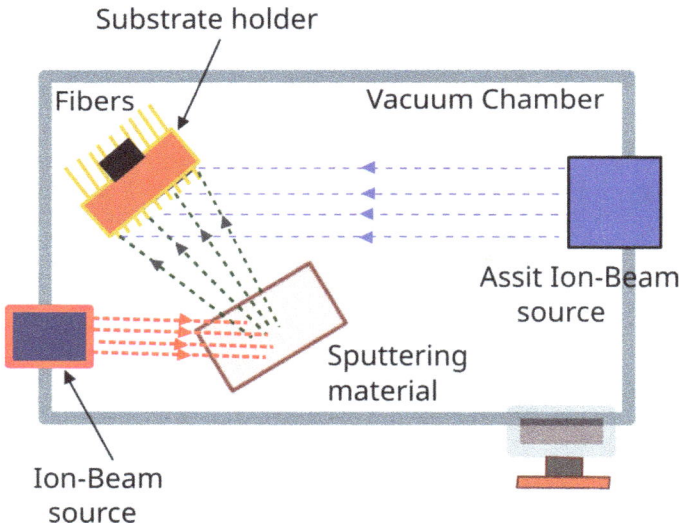

Figure 3.14: Schematic principle for ion-beam sputtering used for coating a dielectric Bragg mirror on a glass substrate. Here, fibers are coated by mounting them on a holder such that the fiber-tips are facing towards the sputtering flux.

material, once ejected, is then deposited onto the substrate. In some cases, an additional second-ion beam is employed to further control the coating properties on the substrate material; see Figure 3.14. The second ion beam serves multiple purposes in the coating process. First, it can be utilized to preclean the substrate surface, removing any organic contamination or impurities that may hinder the coating quality. This ensures a clean and optimal surface for the subsequent deposition process. Additionally, the second ion beam can assist in surface smoothing, refining the substrate's surface texture to enhance the overall coating performance.

The ion-beam sputtering process requires only a low level of vacuum ($< 10^{-4}$ millibar) and also the temperatures are typically below 100 °C. This is advantageous as it simplifies the equipment requirements and reduces the complexity and cost of the coating setup. Moderate operating temperature ensures that the substrate and any delicate components are not subjected to excessive thermal stress during the coating procedure.

Depending on the desired wavelength and the reflectivity of the mirrors, typically the most commonly used coating materials are Ta_2O_5, TiO_2, HfO_2, ZrO_2, Al_2O_3, and SiO_2. In more demanding experiments, where the cavities are limited by thermal noises within the coating, a crystalline coating made of gallium arsenide and aluminum gallium arsenide is utilized. These specific coatings are commonly employed in metrology applications where lowest thermal noise performance are desired, e. g., for high-stability atomic clocks and other precision instruments like gravitational wave detectors.

3.5.1 Annealing of mirrors: enhancing reflectivity

Dielectric mirrors used in high finesse cavities often consist of a glass substrate coated with dielectric coating layers. The commonly used coating materials are Ta_2O_5 and SiO_2. Ta_2O_5 are particularly one of the most important materials for coating due to its high refractive index and excellent resistance to damage caused by incident power. However, a notable issue arises during the deposition process, wherein the coating layers tend to become substoichiometric. To address this concern and to further improve the mirror's reflectivity, a crucial step involves annealing, which entails heating the mirror to a high temperature after the coating process. Therefore, fiber mirrors are annealed before assembling the fiber Fabry–Perot cavity for improved finesse performance, as reported in [2]. Annealing allows the homogenization of the oxide dielectric layers and reduces the coating losses by a few tens of ppm.

3.6 Assembling of a fiber cavity

Assembling a fiber cavity involves achieving optimal alignment between the two fiber mirrors to maximize light transmission through the cavity when it is in resonance. Once

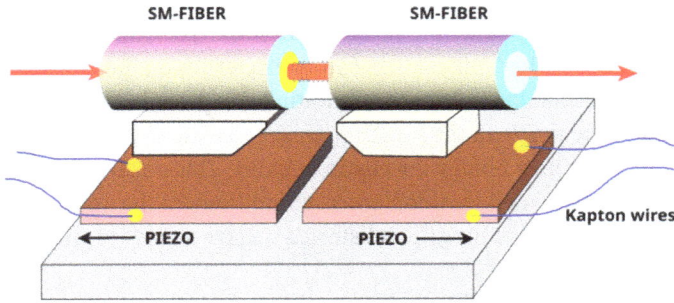

Figure 3.15: Figure (a) shows a specific assembly example of FFPC used in [6]. Figure (b) is the picture of the assembled cavity.

the alignment is established, it is crucial to secure the fiber mirrors on tunable length actuators, such as piezoelectric devices. This enables precise adjustment of the cavity's resonance frequency. Additionally, the entire cavity system should be securely mounted on a stable base to maintain alignment over time (see Figure 3.15).

The specific assembly process for a Fiber Fabry–Perot Cavity (FFPC) may vary depending on the fiber-mirror configuration and the intended application. Here, we describe the assembly process for an FFPC with two concave fiber mirrors. In this configuration, the cavity waist coincides with the cavity center, allowing for open and wide optical access to atoms located at the cavity center. This type of FFPC arrangement has been utilized for interfacing atoms and photons through the cavity [6].

By carefully following the assembly steps, the FFPC can be precisely constructed, ensuring optimal light coupling efficiency and efficient light transmission into the cavity. The following steps are implemented to assemble an FFPC.

- *Splicing:* The opposite end of each fiber mirror is spliced to a fiber pigtail, which enables light coupling in and out of the cavity via a fiber collimator. Commercial fusion fiber-splicing machines are used to achieve low-loss splicing of the fibers.
- *Fiber holder assembly:* Two V-groove holder assemblies are prepared for placing the fiber mirrors. The V-groove holder is a small ceramic piece with grooves engraved on it for placing the fiber. The V-groove piece is glued on top of a shear piezo. Electrical connections to the piezo are made by insulated vacuum compatible Kapton wires. These wires are attached to the piezo using conductive glue. The glue is cured thermally to reduce degassing. Another small ceramic plate is glued under the shear piezo to fix the whole piece on another platform, e. g., on the translation stage.
- *Translation stages:* The holder assemblies are fixed to two independent precision translation stages, and the two fiber mirrors are placed on them. The two fiber mirrors are brought close at a distance of a few tens to a few hundreds of micrometers.
- *Mirror alignment:* The mirror alignment begins with adjusting the fibers in 3D translation and two-dimensional rotation with respect to each other. Two microscope

cameras are used for observing the fiber tips for the orientation and the distance between them along two orthogonal axes. The fiber tips are adjusted to be aligned along the fiber axis and to have a distance close to the desired values. The observables for optimizing the cavity alignment are the transmission and the reflection from the cavity. The cavity length is scanned using piezo of the translation stage. The optical setup used for probing the cavity line shape is described in Section 3.8.

– *Gluing of fiber mirrors:* The ideal alignment would give a maximum transmission of a laser beam at the cavity resonance. In this position, fibers can be glued to the holders. In the final stage, the two holders can also be fixed (glued) to a common base plate while keeping the sensitive alignment of the fiber mirrors. The ultraviolet curable glues with very low degassing for vacuum application can be used for this purpose. Controlled and slow hardening of the UV-glue can be achieved by controlling the UV-radiation intensity for curing.

3.6.1 Ferrule-based fiber cavity

It is worth mentioning that in reference [22], an easier FFPC assembly is demonstrated, which can be used for the spectroscopy of gases [23]. This implementation utilizes glass ferrules that have a bore drilled to guide the two fiber mirrors in near alignment along the fiber axis, as depicted in Figure 3.16. The bore diameters are similar to the fiber diameters. The assembly process involves introducing the fibers into the bore and ad-

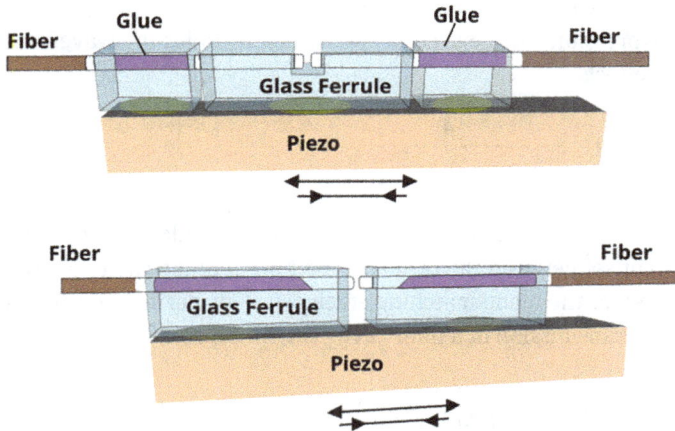

Glass ferrule based Fiber Fabry-Perot Cavity

Figure 3.16: In [22], glass ferrules are used for assembling fiber-based cavities. Two fibers, laser machined and dielectric coated at the fiber ends, are introduced in a segmented glass ferrule, which is glued to the piezo. This figure is adapted from [22] with permission © The Optical Society.

justing the distance between the mirrors to the desired value. Subsequently, the fibers are securely glued inside the ferrules. A small amount of low-viscosity glue is inserted into the bore. The surface tension force allows the glue to spread uniformly around the fibers and within the ferrule. The glue is then cured using UV light, resulting in a stable configuration of the fibers. For the resonance tuning of the cavity, a shear piezo is glued either to a single or segmented ferrule depending upon the desired length tuning of the cavity. It has been shown that this ferrule-based configuration is intrinsically very stable and can be used for the spectroscopy of gases by observing the shift and broadening of the cavity resonance line even during the scan mode [23].

3.7 Losses in fiber cavities

Fiber cavities consist of two fiber mirrors directly made on the fiber tips. Therefore, the propagation mode of the fiber is directly coupled to the mode of the cavity. As the fiber mode is different from the cavity mode, there is only a partial overlap between these modes. The mode mismatch between the fiber and the cavity gives rise to cavity transmission losses, which depend on the resonator's length. For shorter resonators with a cavity length of few 100 µm, the mode mismatch is small, as can be calculated from the following expression of mode matching efficiency (η):

$$\eta = \left(\frac{2\omega_f \omega_m}{\omega_f{}^2 \omega_m{}^2} \right)^2 \tag{3.10}$$

where ω_m and ω_f are the waists of the cavity modes on the mirror and the fiber, respectively. A more complete expression was derived in [10], which includes the wavefront curvature and the lensing effect,

$$\eta = \frac{4}{(\frac{\omega_f}{\omega_m} + \frac{\omega_m}{\omega_f})^2 + (\frac{\pi n \omega_f \omega_m}{\lambda_R})}, \tag{3.11}$$

where ω_f and ω_m are the fiber mode's beam waist and the cavity mode's waist on the fiber mirror. If there is an offset between the center of the fiber and the cavity mirror, additional losses further reduce the mode-matching efficiency. Considering the various factors, the total resonance transmission of a fiber cavity can be expressed as follows:

$$T_{\text{eff}} = \eta_1 \eta_2 \eta_f \left(\frac{T}{T + L} \right)^2, \tag{3.12}$$

where η_1, η_2, η_f are the mode matching efficiency at the two cavity mirrors and the fiber coupling efficiency to the cavity mode, respectively.

3.8 Optical characteristics of a fiber resonator

The optical characteristics of an FFPC can be investigated by recording the transmission and reflection profiles using a laser as a probe. Either the laser or the cavity can be scanned across a cavity resonance. As FFPCs have lengths of a few tens to a few hundred micrometers resulting in a free spectral range of a few terahertz, it is easier to scan the cavity length to observe two consecutive longitudinal cavity modes separated by one free spectral range. The optical layout of the cavity characterization setup is shown in Figure 3.17. A probe laser beam travels through an optical isolator to avoid back reflection to the laser. An EOM in the beam path modulates the phase of the laser light and creates frequency sidebands at the rf-driving frequency. These sidebands can be used to lock the cavity length using the Pound–Drever–Hall technique (see Chapter 2). Also, in the scanning mode, the sideband can be used as a frequency marker for the scans. As shown in Figure 3.17, the laser beam is coupled in and out of the cavity using fiber collimators and incident on two photodiodes. Polarization optics having a polarizer, half-wave plate, and the quarter-wave plate allows specific input polarization into the cavity.

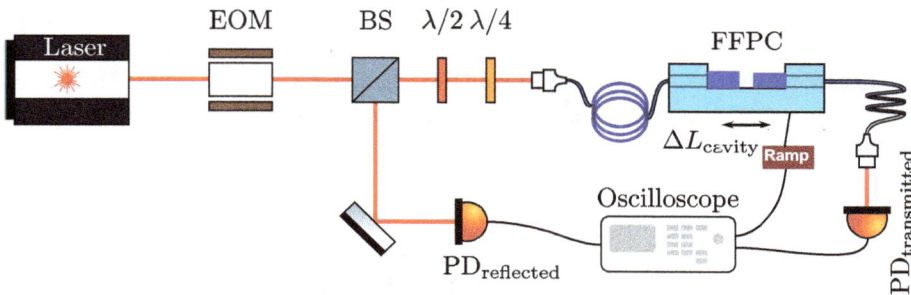

Figure 3.17: A laser is phase modulated using an EOM to generate two frequency sidebands at 250 MHz and coupled to the fiber of an FFPC. A set of polarization optics keeps the fiber polarization close to circular on the input fiber mirror.

As shown in Figure 3.18, the cavity transmission spectrum has a Lorentzian line shape. The linewidth of the cavity is measured by estimating the full-width, half-maximum (Δv) of the transmission (or reflection) dip with respect to the EOM sideband as a frequency marker [22]. Finesse can be estimated from the FSR (= $c/2L$), a geometrical parameter, and the measured linewidth of the cavity Δv:

$$F = \mathrm{FSR}/\Delta v \tag{3.13}$$

The reflection spectrum of an FFPC shows an asymmetry along the wings of the Lorentzian, which is a typical feature in a fiber-coupled resonator. This asymmetric

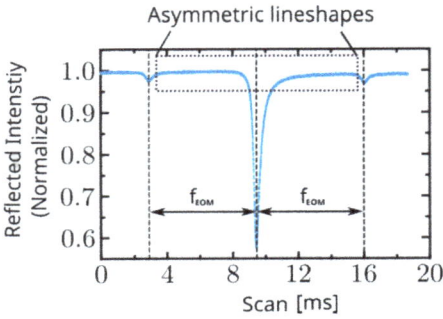

Figure 3.18: Reflection spectrum of a fiber Fabry–Perot cavity. The asymmetric shape along the wings is a specific feature arising due to spatial filtering of the reflected light and interference effect. The overall spectrum can be fitted by a sum of Lorentzian and a dispersive function.

shape can be very well approximated by the sum of a Lorentzian function and its corresponding dispersive function [6]. A brief explanation for this is given below.

Asymmetric line shapes

In contrast to free-space cavities, fiber cavities have severe constraints on the mirror geometry (radius of curvature, mirror diameter) and the cavity length. The losses increase due to a mode mismatch between the mode-field diameter of the fiber-guided mode and the cavity mode for the longer fiber cavities. For cavity lengths beyond hundreds of micrometers, the size of the beam waist on the fiber mirror becomes more significant, and the mode overlap between the cavity and fiber reduces. In the transmission spectrum of the cavity, mode mismatch produces a reduced resonance peak height. However, a more dramatic effect of mode matching is seen in the reflection spectrum of an FFPC. This reflects in terms of deviation from a simple Lorentzian profile close to the cavity resonance. The mode-filtering action and the interference effects at the input fiber produce dispersive effects on the reflected light [1], Figure 3.18. The input fiber, apart from the injected light into the cavity, also guides the reflected light from the cavity and acts as a mode filter for this light. For a nonperfect, mode matching, the interference of the intracavity field with the promptly reflected mode-mismatched field exhibits the sum of a Lorentzian and a dispersive function.

3.8.0.1 Fiber cavities with mode matching optics

When designing fiber cavities with lengths approaching the millimeter range, it is crucial to address the significant mismatch between the propagating mode of the fiber and the mode of the cavity. Clipping losses, which become increasingly significant as the cavity length increases (equation (3.10)), pose a challenge in achieving efficient mode matching. Reference [8] introduces an elaborate technique for mode shaping the fiber light field to overcome these challenges. The use of special fibers, such as gradient-index

(GRIN) fibers, allows for shaping the output mode of a single-mode (SM) fiber by converging or diverging the light as it exits the fiber. This technique requires precise cutting of the GRIN fiber length and careful splicing of two different types of fibers while maintaining the single-mode Gaussian output.

Figure 3.19 illustrates the concept of attaching a single-mode fiber with a 100-micrometer GRIN fiber to shape the light mode for a 500-micrometer cavity length. By integrating this assembly into the cavity, both the field diameter of the fiber and the wavefront curvature can be precisely matched with the cavity mode. It is recommended to use a large-core GRIN fiber compared to the SM fiber to avoid distortions at the core-cladding interface. The gradient-index profile of the GRIN fiber determines its convergence or divergence properties for different lengths. By performing wave-optical simulations of light propagation, it is possible to predetermine the required length of the GRIN fiber for a specific cavity length.

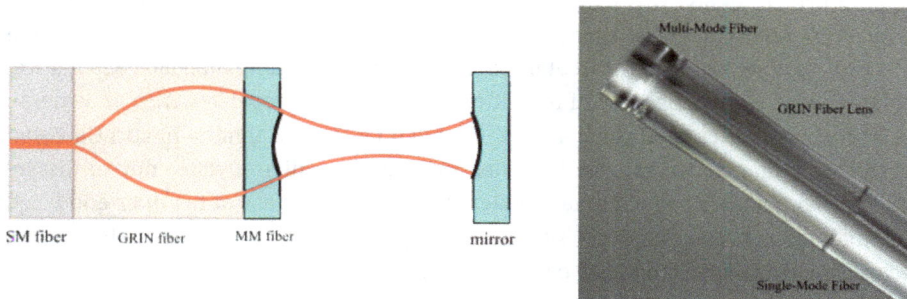

Figure 3.19: Splicing a GRIN fiber of appropriate length after the single mode fiber allows mode matching of the fiber propagation mode to the cavity mode. This arrangement is particularly used for longer cavities.

This mode-shaping technique using GRIN fibers offers a solution to the mode mismatch challenges in fiber cavities with longer lengths. By carefully designing and integrating these fiber components, efficient mode matching can be achieved, reducing clipping losses and optimizing the performance of the cavity for various applications.

3.8.1 Multishoot technique for longer fiber cavities

In many applications such as spectroscopic sensing, fiber-based frequency filters, and the integration of ion traps with cavities; there is a strong demand for longer fiber cavities with cavity lengths approaching a few millimeters. The extended length of the cavity offers several significant benefits.

One advantage is that a longer cavity length results in a smaller free spectral range. The free spectral range refers to the separation between adjacent resonant frequencies in the cavity. By reducing this range, finer frequency resolution can be achieved, which is crucial in spectroscopic sensing and frequency filtering applications.

Additionally, a longer cavity allows for a larger distance between the mirror surface and the medium, particularly in cases where trapped ions are involved. This increased distance is essential for preventing unwanted discharge of the ions when they are in close proximity to the dielectric mirror surface. It ensures a safe and stable environment for the trapped ions within the cavity. However, as the cavity length increases, the clipping losses or losses due to mode mismatch between the fiber and the cavity mode increase. To overcome the clipping losses, fiber mirrors with a larger diameter and radius of curvature are required [17].

For fiber mirrors with a larger diameter, the size of the mirror can significantly extend up to the fiber edges. However, creating a mirror surface extending over the whole fiber end is substantially challenging using the laser ablation process. First, the heat transport toward the edges of the fiber cladding is inefficient, resulting in nonuniform temperature gradients. This makes it difficult to extend mirror surfaces over the majority of the fiber-end facet. Additionally, different doping concentrations between the core and cladding can lead to circular ridges at the core-cladding interface when a single laser pulse is used, because of the different absorption properties. Also, in general, the mirror profiles created using a single-shot CO_2 laser pulse tend to have a Gaussian depression rather than a spherical profile, resulting in smaller effective mirror diameters even with a laser beam diameter extending over the surface of the fiber-end facet. The small mirror sizes can cause significant clipping losses at the longer FFPC lengths because the transverse mode size becomes comparable to the effective mirror diameter.

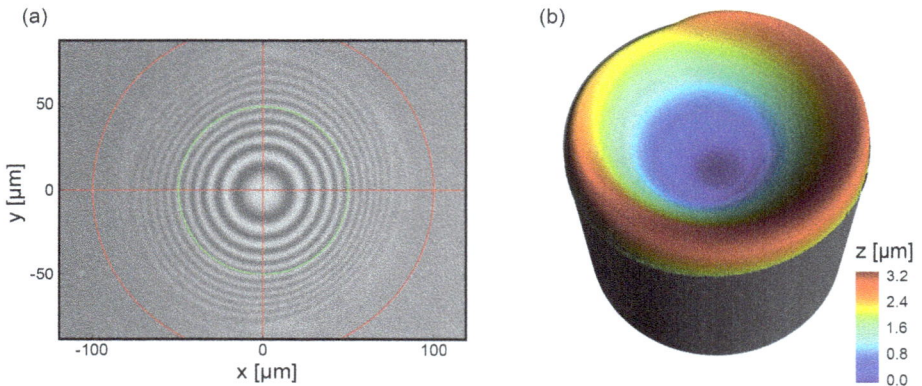

Figure 3.20: This figure shows an example of large spherical surface machined by the CO_2 laser. A special SM fiber with 200 μm diameter is used in [17]. Figure (a) shows the interferogram of the laser machined surface. The red circle shows the initial fiber diameter, the green circle shows the area over which the structure was optimized. The cross-hairs indicate the center of the fiber. Figure (b) is the reconstructed surface using phase-shifting interferometry. This figure is reprinted with permission from [17] © The Optical Society.

The dot-milling technique, as demonstrated in [17], involves using multiple weak laser pulses in a predefined position on the fiber such that the effective surface profile is a very smooth spherical surface with a much larger radius of curvatures for the mirror. Spherical mirror structures with excellent quality and shape deviation with a few tenths of the wavelength of light can be generated with targeted points milling on the surface. Figure 3.20 shows a CO_2 laser pulse pattern on the fiber surface and the resulting mirror structure.

3.8.2 Polarization mode splitting

Polarization mode splitting is a phenomenon observed in optical cavities where a single resonant mode splits into two distinct modes with different polarization states. This occurs when the cavity supports multiple polarization eigenmodes that have slightly different resonant frequencies.

There are several factors that can contribute to polarization mode splitting in an optical cavity. These include birefringence in the cavity material, stress-induced birefringence, or nonuniform stress or strain within the cavity structure. These factors result in a polarization-dependent refractive index, causing a frequency difference between the two orthogonal polarization modes.

When light is injected into the optical cavity, it excites both polarization modes. However, due to the frequency difference between the modes, the phases of the two polarization components accumulate at different rates as the light undergoes multiple reflections within the cavity. This phase difference causes the initially degenerate resonant mode to split into two separate modes with distinct resonant frequencies. The following expression gives the frequency difference between the two orthogonal polarization modes of the cavity [13]:

$$\frac{\Delta\nu_{\text{pol}}}{\Delta\nu_{\text{FWHM}}} = \frac{F}{2\pi}\Delta\phi \tag{3.14}$$

Polarization mode splitting is particularly pronounced in high-finesse cavities due to the multiple reflections between the cavity mirrors, which enhance the effect of phase accumulation. The two main reasons for the polarization splitting in fiber cavities can be the following:

1) Birefringence: This can arise due to several factors such as birefringence in the cavity material, stress-induced birefringence, or the presence of nonuniform stress or strain within the cavity structure. Birefringence leads to a polarization-dependent refractive index. The light circulating inside the cavity penetrates the coating within a tiny fraction of the coating thickness. However, the different effective refractive indices of the light experienced along the two orthogonal axes create polarization-dependent penetration depth and different phase shifts.

2) Ellipticity in the fiber mirrors: For fiber mirrors with ellipticity, the frequency splitting of the two polarization modes depends on the radius of curvature of the mirrors along the polarization direction. The difference in radius of curvature between the two orthogonal axes induces a polarization-dependent phase shift, and consequently, polarization mode splitting of the cavity modes. This frequency splitting is because the cavity resonance frequency for a linear polarization mode depends on the mirror's radius of curvature along the polarization direction. Therefore, the different radius of curvature along two orthogonal axes of a mirror induces polarization-dependent phase shift, and consequently, the polarization mode splitting of the cavity modes [28]:

$$\Delta\phi = \frac{\lambda}{2\pi}\left(\frac{R_a - R_b}{R_a R_b}\right). \tag{3.15}$$

From equation (3.14), it is clear that this effect can produce polarization mode splitting of the order of hundreds of MHz for fiber-based Fabry–Perot resonators.

Mitigating the polarization mode splitting involves compensating the phase shift induced by one mirror from the other one. The reduction of polarization splitting is possible by rotating the fiber mirrors with respect to each other. If the phase shift induced by the two mirrors is $\Delta\phi_1$ and $\Delta\phi_2$, then the effective round trip phase shift, when there is an angle θ between the axis of the two mirrors, is given by [28],

$$\Delta\Phi_{\text{eff}} = \sqrt{(\Delta\Phi_1{}^2 + \Delta\Phi_2{}^2 + 2\Delta\Phi_1\Delta\Phi_2 \cos 2\theta)} \tag{3.16}$$

For an adequately chosen angle between the fiber axis, the effective phase shift can be reduced down to the difference of the individual phase shift of the two mirrors. And this could be perfectly canceled for two mirrors having the same ellipticity.

Experimentally, one of the fiber mirrors is placed on a rotatable fiber v-groove or holder. During the alignment of the two mirrors for making a cavity, one of the mirrors is rotated with respect to the other mirror, and the polarization mode splitting is reduced for the best alignment. Figure 3.21 shows the setup used for minimizing the polarization mode splitting.

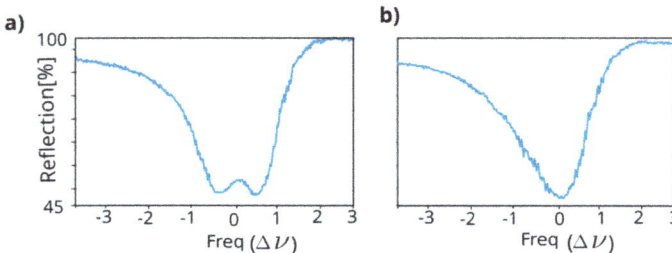

Figure 3.21: Figure (a) shows the reflected spectrum of a fiber cavity when the fiber mirrors are rotationally oriented to have maximum ellipticity axis orthogonal, producing the maximum polarization splitting of the resonance modes. Figure (b) is the reflection spectrum with minimal polarization splitting of eigenmodes.

3.9 Nanofiber and nanofiber resonator

Nanofibers and nanofiber resonators are fascinating areas of research in the field of photonics sensors, quantum optics, and quantum technology. In this section, we will first introduce the optical nanofiber systems and then explore the techniques, which are employed for creating reflectors on nanofiber systems to produce nanofiber resonators.

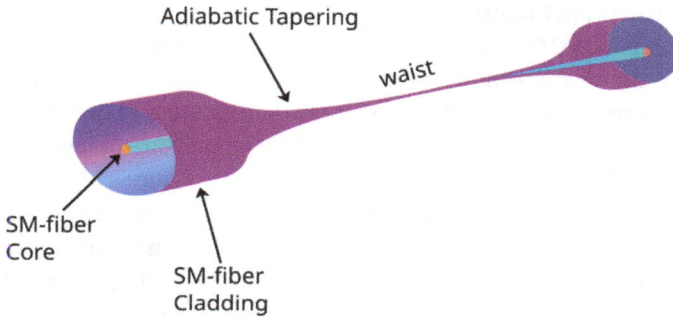

Figure 3.22: A nanofiber system obtained by tapering a commercial single-mode fiber into biconical tapers with a nanofiber waist at the center.

Optical nanofibers (and microfibers) are custom-modified commercial optical fibers with biconical tapers leading to a waist size of a few hundred nanometers (close to 1 µm). As shown in Figure 3.22, these optical systems have three distinct light propagation regions

- *Untapered single-mode fiber regions*: These regions are the standard and unmodified single-mode fibers and exist at both the input and output sides of the nanofiber. Here, light propagates through core-cladding guided modes in the weak guiding regime, as explained for the SM-fibers in Chapter 1.
- *Tapered fiber regions*: These regions involve a gradual thinning of the single-mode fiber (SM-fiber) until it reaches the nanofiber waist at the center. The electric field of light in the tapered regions gradually gets compressed in the radial direction until the light is guided via the core and the surrounding air/medium interface. Adiabatic tapering is essential to ensure maximum coupling of the SM-fiber propagation from the input to the output side.
- *Nanofiber waist region*: In this region, the fiber diameter remains minimal and relatively constant. Light is guided through cladding-air-guided modes with strong radial confinement. Most of the light energy propagates outside the fiber as an evanescent wave.

Tapered nanofibers are produced by controlled axial pulling of a standard silica fiber while heating a small portion of the fiber close to the melting point. The localized and controlled heating mechanism is required to raise the glass temperature to the optimum value. Highly controlled and precise pulling is essential to allow adiabatic tapering resulting in a very smooth and ultralow loss connection to the main fiber. As the light travels outside the fiber cladding region, close to the waist, the strongly confined evanescent field in the nanofibers can be used for strong light-matter interaction. This interaction is enhanced by placing matter system in close proximity to the fiber waist. The coupling of light to the surrounding medium via evanescent coupling makes nanofiber a perfect candidate for various sensing applications.

The light propagation outside the fiber in the tapered region makes the light coupling also very sensitive to any dust or contamination and can dramatically compromise the performance of the nanofiber system. Therefore, the whole process of nanofiber production takes place in a very clean environment. The following steps are followed to taper the standard single-mode fibers down to the desired waist; see Figure 3.23.

–   First, the fiber's polymer-coated (or metal coating) protection is removed with a fiber stripper (or chemical process) and cleaned thereafter using isopropyl alcohol. The fiber is then clamped on two opposing high-precision translation stages by a pair of rotatable fiber clamps to ensure straight and torsion-free clamping. The length of the bare fiber can be a few centimeters depending on the tapered length and the waist.

–   The bare-fiber region is imaged on a microscope camera for any contamination, which may require cleaning the fiber again.

–   A high-purity hydrogen-oxygen flame (or any other heating mechanism, e. g., electric arc, CO_2 laser beam, or a ceramic microheater) is used to attain a temperature close to the melting point of the glass. The desired temperature is when the glass becomes very soft but does not yet melt.

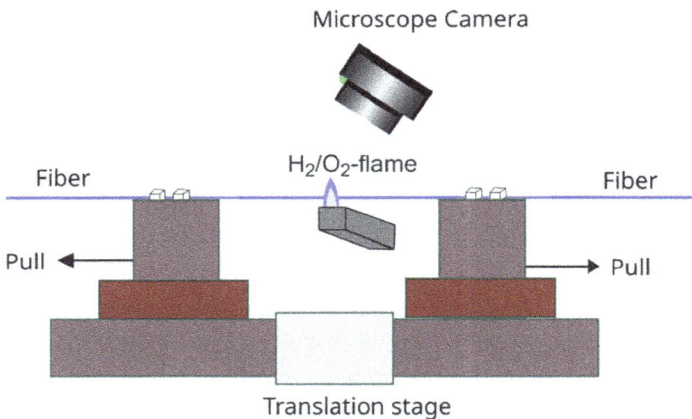

Figure 3.23: Schematics of a fiber tapering setup. A bare single-mode fiber is fixed at two points that are a few mm apart. Flame heating is applied while the fiber ends are slowly pulled in the opposite direction. This process produces a nanofiber.

- Fibers are then pulled across the flame on the opposing end. The high precision and computer-controlled translation stages allow submicron positioning and control of the pulling process. This is required to create a very smooth and low-loss tapered-optical nanofiber system. Depending on the desired waist (several micrometers to a few hundred nanometers) and the length of the tapered region, fiber position with respect to the flame can also be changed. The adiabaticity condition for tapering (Section 2.4.1) can be maintained with the optimum parameters for fiber pulling. The above includes the temperature and position of the heating flame and the rate of fiber pulling, which are empirically found and automated via computer.
- Laser transmission is continuously monitored during the pulling and fabrication process. Due to changes in the fiber geometry in this process, the fiber first loses its single-mode light propagation. The interference between various fiber modes gives rise to oscillations in the transmission profile. As the fiber-pulling process continues, the fiber system becomes single mode again, and the transmission becomes maximum again. At this point, the pulling process is stopped (see Figure 3.24).

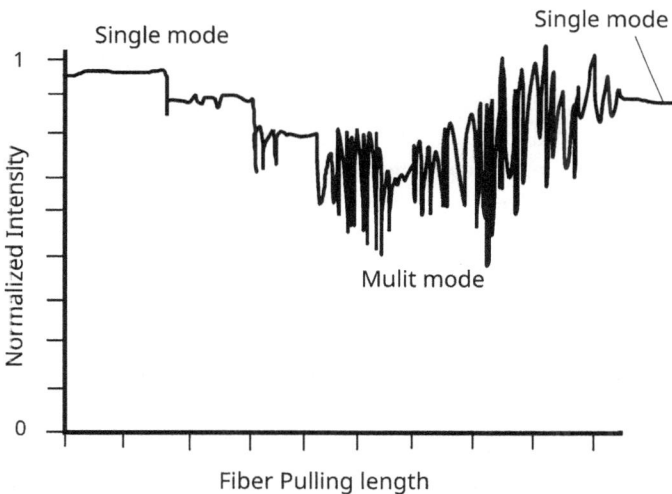

Figure 3.24: The transmission of a single-mode fiber changes as the fiber is heated and pulled for tapering down to nanofiber waists. In the beginning and at the end of the process, fiber behaves as a purely single-mode transmission device.

3.9.1 Adiabaticity condition for nanofiber taper

The adiabaticity condition for nanofiber tapering is an important criterion that ensures low losses and high transmission efficiency through a nanofiber system. The adiabatic tapering down to the nanofiber waist allows the core-guided mode to adiabatically escape to the cladding and finally to the evanescent wave in the waist region. Qualitatively, one should follow the following condition to allow for low losses and high single-mode transmission efficiencies.

The slowness of the pulling should follow the criterion introduced by Synder and Love [24, 20]:

$$\left|\frac{d\rho}{dz}\right| \ll \frac{\rho}{z_b}; \quad \text{where } z_b = \frac{2\pi}{\beta_1 - \beta_2}, \quad (3.17)$$

where $|\frac{d\rho}{dz}|$ is the rate of change of the fiber radius with respect to the axial position, ρ is the fiber radius, and z_b is the beat length between the two modes involved in the tapering process. The beat length is defined as the distance over which the phase difference between the two modes (HE_{11} and HE_{12}) accumulates by 2π and is given in terms of the difference between the propagation constants for the two modes, which are β_1 and β_2.

The adiabaticity condition ensures that the change in fiber radius occurs gradually and smoothly, allowing the guided modes to adjust and maintain their fundamental characteristics during the tapering process. If the tapering rate is too fast or the condition is not satisfied, the guided modes can undergo significant mode mixing, resulting in increased scattering and higher losses. To achieve adiabatic tapering, precise control of the tapering parameters is necessary. This includes controlling the heating and pulling mechanisms to ensure a gradual reduction in the fiber diameter. By carefully controlling the tapering rate, it is possible to achieve a smooth and low-loss transition from a standard fiber to a nanofiber with a significantly reduced diameter.

3.9.2 Nanofiber resonators

Nanofibers offer a wide range of applications due to their exceptional light confinement near the waist region. The strong evanescent coupling of light enables robust light-matter interactions in close proximity to the fiber. Further enhancement of these interactions can be obtained by combining a nanofiber's strong light confinement with an optical resonator's multipass effect. In [30], authors have demonstrated the integration of the fiber Bragg mirrors-based optical cavity with the nanofiber platform. In the context of nanofiber resonators, fiber-integrated Bragg mirrors are used to create an optical cavity that enhances the interaction between light and matter. The nanofiber itself acts as the waveguide, guiding light along its length, while the Bragg mirrors provide the necessary reflectivity to form a resonant cavity. The ability to selectively filter light by combining the cavity with the nanofiber allows resonant enhancement within a specific wavelength. Moreover, the integration of Bragg mirrors simplifies the alignment and coupling of the nanofiber resonator with other optical components, making it compatible with fiber-based systems. This compatibility paves the way for seamless integration into larger quantum networks and facilitates the deployment of nanofiber resonators in advanced sensing setups.

3.9.3 Fiber integrated Bragg mirrors

Fiber-integrated Bragg mirrors are optical devices that use the principle of Bragg reflection to selectively reflect specific wavelengths of light. These mirrors are designed and integrated directly into optical fibers, allowing for compact and efficient optical components with tailored reflection properties. The Bragg mirror structure consists of alternating layers of materials with different refractive indices. Each layer has a thickness equal to half the desired wavelength of light to be reflected. When light propagates through the Bragg mirror, the constructive interference between the reflected waves from each layer results in strong reflection at the desired wavelength while transmitting other wavelengths. In the case of fiber-integrated Bragg mirrors, the Bragg mirror structure is directly formed on or within the optical fiber itself. This integration allows for seamless integration of the mirror into the fiber optic system, reducing the need for additional bulk components and facilitating fiber-based applications.

Fiber Bragg mirror fabrication techniques can be used to implement a cavity within a nanofiber system [29, 21, 9]. This technique combines the advantage of strong confinement of the evanescent field with the multipass effect of the cavity. Here, we present a short description of the methods used to create fiber Bragg mirrors as demonstrated in [30].

Two mirrors for the nanofiber resonators are realized by either creating nanophotonic structures on the waist region or by creating fiber Bragg grating in the untapered part of the nanofibers, Figure 3.25.

– Nanophotonic-based mirrors are realized by nanofiber Bragg gratings. These gratings are fabricated by using a focused ion-beam milling technique. A series of equidistant slits are created to produce a periodic modulation of the effective refractive index. The typical distance between the slits is a few hundred nm. Suppose these slits are created at the waist region of the tapered nano-fiber. In that case, one can create two mirrors (fiber Bragg grating reflectors) required for an optical resonator. Depending on the wavelength of light, one can produce precalculated structures using a very precise ion-beam milling technique. These commercial ion-beam millings are performed under vacuum and require precise positioning and translation of the tapered nanofiber with respect to the ion beam. Also, this process requires careful discharging of the insulating silica glass while milling using the charged beam. Overall, precise handling the nanofiber with the ion-beam milling technique and analyzing the created grating structures is a complex process.

– Another method of generating mirrors for nanofiber resonators is using the techniques of writing fiber Bragg gratings.

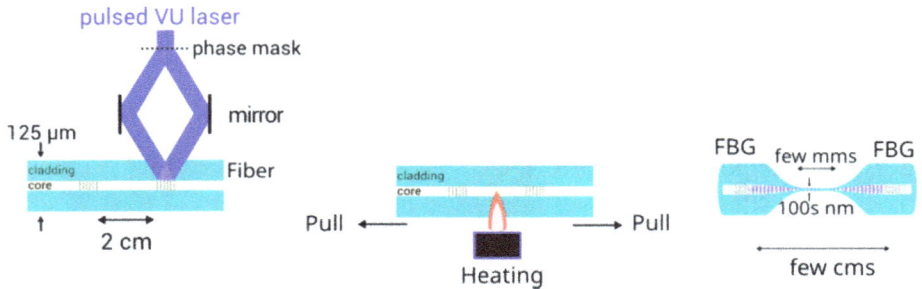

Figure 3.25: An example of Bragg mirror inscription on a fiber using a UV-light illumination. A phase-grating produces the desired spatial intensity modulation for the UV light.

In this approach, fiber Bragg gratings are written in the untapered fiber section of the nanofiber. As shown in Figure 3.25, two fiber Bragg mirrors are engraved on the two sides of the tapered optical fiber. These two fiber mirrors constitute the optical mirrors for the nanofiber resonator. Fiber Bragg mirror can be created by standard techniques of laser writing used in optical fiber technology. When fused silica is illuminated by very high optical intensity, its refractive index can be permanently changed. Therefore, using interferometric techniques, certain kinds of intensity-modulating patterns can generate periodic refractive index modulation in the fibers. Typically, pulsed lasers in Ultraviolet (UV) wavelength and with periodic intensity profiles are used to create refractive index modulation at the core region of the fiber. The laser beam is split and interferometrically combined at an angle to create desired intensity modulation pattern. Fibers are loaded with hydrogen to make the laser inscription more efficient.

For creating nanofiber resonators, a standard fiber is first written with two fiber Bragg mirrors at a few cm distances. After this, the region between the mirrors is heated and pulled with ultrahigh precision to create a tapered fiber between the mirrors. As described in Section 3.9.1, the waist of the tapered profile is optimized to have the highest transmission for a laser beam injected into the fiber system.

3.9.4 Whispering gallery mode resonator

In an optical Whispering Gallery Mode (WGM) resonator, light waves circulate along the perimeter of the curved structure, bouncing off the inner surface with each revolution. This results in a long optical path length and enables the resonator to trap light with high efficiency. The light waves are confined due to the phenomenon of total internal reflection at the curved surface, which prevents them from escaping the resonator.

The WGM resonator can take various forms [27], including microspheres, microdisks, microrings, or even tapered fibers [18]. These structures are typically made of materials with high optical transparency, such as glass or crystalline materials. The size of the resonator determines the resonant frequencies and the characteristics of the supported whispering gallery modes.

WGM resonators have one of the highest quality factors and the smallest mode volume. The high quality factor for the WGM resonator indicates low optical losses and enhanced light confinement within the resonator, enabling long photon lifetimes and high sensitivity to changes in the surrounding environment (i. e., high sensitivity sensor). The light coupling efficiency in and out of the WGM resonators reaches an efficiency of 100 % using evanescent coupling. Therefore, WGM resonators have found applications in various fields, including optical sensing, nonlinear optics, cavity quantum electrodynamics, and telecommunications. They can be used as sensitive sensors for detecting small changes in temperature, refractive index, or other environmental parameters. Additionally, they play a crucial role in the development of compact and efficient lasers, optical filters, and optical signal processing devices. Here, we discuss the fiber-based bottle resonators and their fabrication process.

Fiber-based WGM resonators can be manufactured using nanofiber manufacturing techniques of precise heating and pulling of standard single-mode fibers. This technique allows for forming a bottle resonator profile that has the following distinct features:

-   The axial profile of a bottle resonator has a parabolic shape with a radius profile as

$$R(z) = R_0\left(1 - \frac{1}{2}(\Delta k.z)^2\right),$$

(3.18)

where R_0 is the central radius and Δk is the radius of curvature (see Figure 3.26).

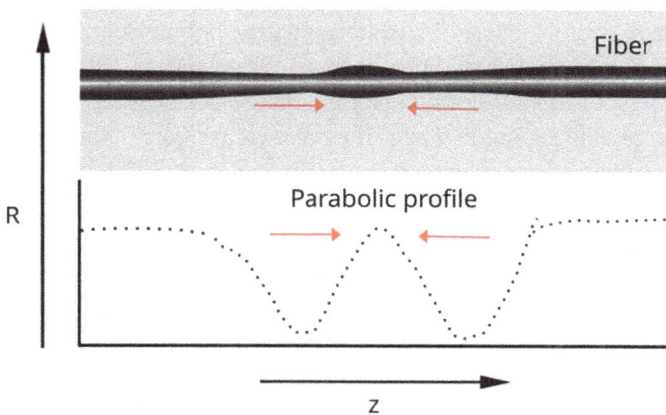

Figure 3.26: Spatial profile of a bottle resonator along the axial direction. The bulge at the center has a parabolic profile.

- The axial profile results in light modes with two pronounced regions of enhanced intensity. These are called "caustics."
- Light field has an effective harmonic potential along the resonator axis between the caustic points defined by the angular momentum barrier (see Figure 3.28).
- The free spectral range of the bottle resonator is defined as the frequency spacing between modes with consecutive quantum numbers along the axial and azimuthal directions. The axial and azimuthal FSRs are given by (see Chapter 1):
FSR along the axial direction is

$$\Delta v_m \approx \frac{c}{2\pi n} \frac{1}{R_0} \tag{3.19}$$

FSR along the azimuthal direction is

$$\Delta v_m \approx \frac{c}{2\pi n} \Delta k \tag{3.20}$$

The resonator radius determines the azimuthal FSR, while the axial FSR depends only on the curvature of the resonator.
- Light can be coupled in and out of the resonator at the caustic points via evanescent coupling using nanofibers.

3.9.5 Fabrication of bottle resonators

The fabrication process of fiber-based bottle resonators shares similarities with nano-fiber fabrication techniques, specifically the heat and pull technique is also applied here to obtain the desired shape of the glass fibers while heating close to the melting point. An alternative fabrication technique, termed as "soften-and-compress" in [16], involves pushing flat cleaved ends of two fibers together while applying continuous heat using an arch discharge during the fusion process in a standard fusion splicer. This fusion process [4], with controlled pressure, results in the formation of a smooth bulge region between the two fibers, resembling a bottle shape. The more conventional fabrication of a bottle resonator typically follows the heating and pulling method, which includes the following steps:

- The first step in manufacturing the WGM resonator involves removing the protective layer of a single-mode fiber cladding. A standard optical SM fiber typically has a cladding diameter ~125 μm and core diameter ~5.6 μm. The cladding is usually protected by either acrylic or a metal coating.
- After cleaning the fiber with isopropyl alcohol, it is placed on two clamps attached to motorized precision translation stages. A microscope camera is used to monitor the change in the fiber profile. The entire setup is kept in an ultraclean environment to avoid contamination of the fiber surfaces from dust and other pollutants.

- The fiber is heated and pulled such that a few millimeter-long sections of the resulting fiber has a diameter corresponding to the desired diameter of the WGM resonator, typically a few tens of micrometer. A uniform diameter along the few millimeter lengths is possible by linearly translating the fiber over the heating flame while pulling it at the same time, Figure 3.23.
- This tapered fiber is further heated and slightly pulled locally on the two sides of a tapered section (a few hundred micrometers apart) to form a bulge. The heating is accomplished by either hydrogen-oxygen flame or CO_2 laser beam. The bulge at the central region of the tapered fiber has a parabolic variation of the fiber diameter along the fiber axis. This bulging fiber part forms the bottle resonator or the WGM resonator, Figure 3.27.

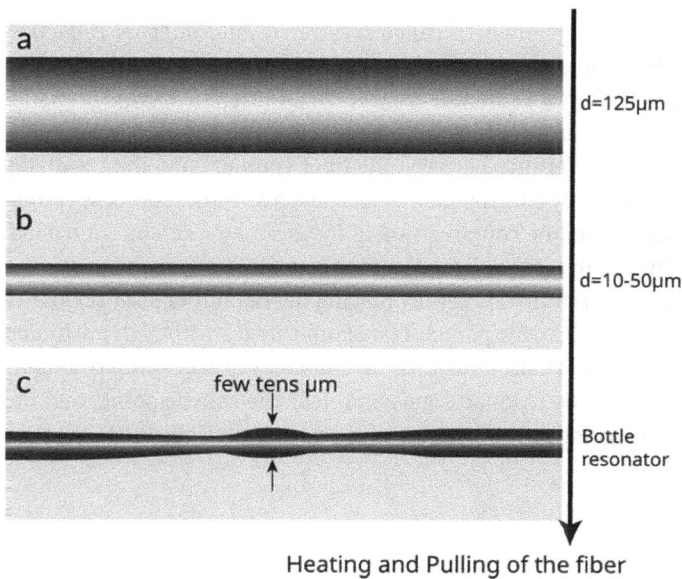

a

d=125μm

b

d=10-50μm

c   few tens μm

Bottle resonator

Heating and Pulling of the fiber

Figure 3.27: A single mode fiber is first tapered down to a smaller waist, and then further heating and pulling at two points create a bulge-shaped prolate profile of a bottle resonator.

- The optical fibers are susceptible and especially fragile at the waist. After fabrication, they are placed and mounted securely on a holder, which can apply mechanical strain along the fiber axis. A piezo at the fiber holding points can provide tuning capability of the resonance frequency of the bottle modes. A shear or bending piezo can be used with a travel range of tens of micrometers.

The light is coupled to the bottle microresonator via evanescent coupling using biconical nanofibers.

3.9.6 Evanescent light coupling into a bottle resonator

WGM resonators exhibit exceptional efficiency when it comes to coupling light in and out of the resonator. The coupling process can achieve nearly 100 % efficiency. In this section, we will explore the concept of evanescent light coupling in the context of a bottle resonator.

The term "evanescent light coupling" refers to the process of transferring light energy from one waveguide to another through evanescent waves. Evanescent waves are electromagnetic waves that decay exponentially away from the waveguide's core. They exist near the surface of the waveguide and can interact with nearby structures or waveguides. To couple light into a bottle resonator, one technique involves utilizing ultrathin tapered optical nanofibers. These nanofibers systems are presented in Section 3.9. When a nanofiber carrying a laser beam is brought close to the surface of the bottle resonator, a fascinating phenomenon called frustrated total internal reflection allows for efficient coupling in and out of the WGM resonator, Figure 3.28. The distance between the nanofiber and the WGM resonator is carefully adjusted to maximize spatial overlap and to match the wave vectors of the evanescent fields. The phase-matching condition ensures constructive interference between the fiber couplers and the resonator fields at the in-coupling junction. By altering the electric field overlaps, such as by changing the distance between the two, the coupling rate of the resonator can be controlled. The following steps outline the process of light coupling into the bottle resonator:

–   The efficient light coupling requires precisely placing the nanofiber's waist close to the surface and at one of the caustic points. Translation stages with high positional resolutions ($\sim 10 - 100$ nm) are used for this. Two degrees of freedom are important, i. e., translation along the resonator axis and near the caustic points and the translation between the nanofiber and bottle resonator to control the mode overlap.

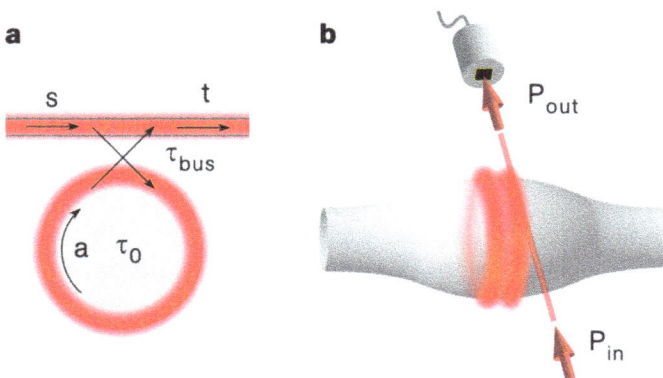

Figure 3.28: Schematic depiction of light coupling from a bottle WGM resonator. A nanofiber bus is used to couple light in and out of the bottle resonator, close to the caustic points. This figure is reprinted from [19] with permission.

The latter requires a much better resolution in position to optimize coupling efficiency by controlling the gap distance. The gap distance has to be controlled within
the decay length of the evanescent fields of the bottle resonator and the coupling
nanofiber.

– Third translation to place the waist of nanofiber with respect to the bottle resonator
requires only micrometer resolution.

– Due to the two caustic points of the bottle resonator, it is possible to simultaneously
access the modes of the bottle resonator with two coupling nanofibers. The caustic
points are typically at least a few micrometers apart and allow placing two coupling
fibers with precision translation stages. The configuration of two coupling fibers on
the bottle resonator is sometimes called an "add-drop" configuration. One of the
fiber coupling lights into the resonator is called the "bus fiber," while the other is
called the "drop fiber." In this configuration, the bottle microresonator can be considered a four-port device.

– For monitoring the amount of light coupled into the bottle resonator while optimization of the coupling efficiency, laser light is coupled to the nanofiber (called bus fiber
in this context). The polarization of light is adjusted to be matched with the polarization of the bottle resonator at the coupling point. Photodiodes are used to monitor
the transmitted light through the coupling nanofibers. As shown in Figure 3.29, the
depth (height) of the photodetector signal for the transmitted light through the drop
(bus) fiber is maximized for the most efficient light coupling.

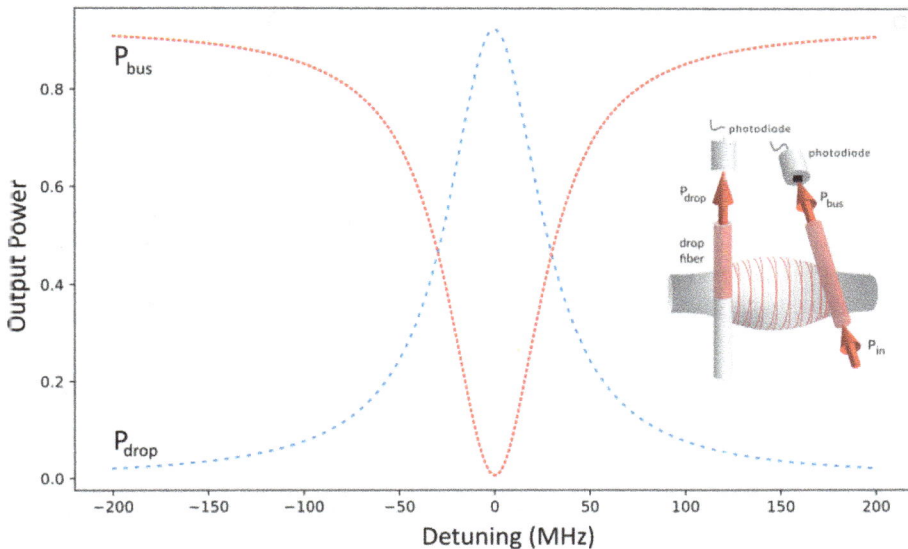

Figure 3.29: The distance and orientation of the coupling fibers with respect to the bottle resonator are
optimized by observing the light transmitted through the fibers (P_{bus} and P_{drop}) anc maximizing the depth
of the resonance peak.

Bibliography

[1]   A. Bick, C. Staarmann, P. Christoph, O. Hellmig, J. Heinze, K. Sengstock, and C. Becker. The role of mode match in fiber cavities. *Review of Scientific Instruments*, 87(1):013102, 01 2016. ISSN 0034-6748. https://doi.org/10.1063/1.4939046.

[2]   B. Brandstätter, A. McClung, K. Schüppert, B. Casabone, K. Friebe, A. Stute, P. O. Schmidt, C. Deutsch, J. Reichel, R. Blatt, and T. E. Northup. Integrated fiber-mirror ion trap for strong ion-cavity coupling. *Review of Scientific Instruments*, 84(12):123104, 12 2013. ISSN 0034-6748. https://doi.org/10.1063/1.4838696.

[3]   Y. Colombe, T. Steinmetz, G. Dubois, F. Linke, D. Hunger, and J. Reichel. Strong atom–field coupling for Bose–Einstein condensates in an optical cavity on a chip. *Nature*, 450(7167):272–276, Nov 2007. ISSN 1476-4687. https://doi.org/10.1038/nature06331.

[4]   Y. Dong, X. Jin, and K. Wang. Selective excitation of high-q resonant modes in a bottle/quasi-cylindrical microresonator. *Optics Communications*, 372:106–112, 2016. ISSN 0030-4018. https://doi.org/10.1016/j.optcom.2016.04.016. URL https://www.sciencedirect.com/science/article/pii/S0030401816302784.

[5]   M. D. Feit and A. M. Rubenchik. Mechanisms of co2 laser mitigation of laser damage growth in fused silica. In *Laser-Induced Damage in Optical Materials: 2002 and 7th International Workshop on Laser Beam and Optics Characterization*, volume 4932, pages 91–102. SPIE, 2003.

[6]   J. C. Gallego Fernández. *Strong coupling between small atomic ensembles and an open fiber cavity*. PhD thesis, Universitäts-und Landesbibliothek Bonn, 2018.

[7]   Y.-S. Ghim and A. Davies. Complete fringe order determination in scanning white-light interferometry using a fourier-based technique. *Applied Optics*, 51(12):1922–1928, 2012.

[8]   G. K. Gulati, H. Takahashi, N. Podoliak, P. Horak, and M. Keller. Fiber cavities with integrated mode matching optics. *Scientific Reports*, 7(1):5556, Jul 2017. ISSN 2045-2322. https://doi.org/10.1038/s41598-017-05729-8.

[9]   J. Hütner, T. Hoinkes, M. Becker, M. Rothhardt, A. Rauschenbeutel, and S. M. Skoff. Nanofiber-based high-q microresonator for cryogenic applications. *Optics Express*, 28(3):3249–3257, Feb 2020. https://doi.org/10.1364/OE.381286. URL https://opg.optica.org/oe/abstract.cfm?URI=oe-28-3-3249.

[10] W. B. Joyce and B. C. DeLoach. Alignment of gaussian beams. *Applied Optics*, 23(23):4187–4196, Dec 1984. https://doi.org/10.1364/AO.23.004187. URL https://opg.optica.org/ao/abstract.cfm?URI=ao-23-23-4187.

[11] R. Kitamura, L. Pilon, and M. Jonasz. Optical constants of silica glass from extreme ultraviolet to far infrared at near room temperature. *Applied Optics*, 46(33):8118–8133, Nov 2007. https://doi.org/10.1364/AO.46.008118. URL https://opg.optica.org/ao/abstract.cfm?URI=ao-46-33-8118.

[12] P. Lehmann, S. Tereschenko, and W. Xie. Fundamental aspects of resolution and precision in vertical scanning white-light interferometry. *Surface Topography: Metrology and Properties*, 4(2):024004, 2016.

[13] T. W. Lynn. *Measurement and control of individual quanta in cavity QED*. PhD thesis, 2003.

[14] M. A. A. Mamun, P. J. Cadusch, T. Katkus, S. Juodkazis, and P. R. Stoddart. Quantifying end-face quality of cleaved fibers: Femtosecond laser versus mechanical scribing. *Optics and Laser Technology*, 141:107111, 2021. ISSN 0030-3992. https://doi.org/10.1016/j.optlastec.2021.107111. URL https://www.sciencedirect.com/science/article/pii/S0030399221001997.

[15] L. Mingzhou. *Developmen of fringe analysis techniques in white light interferometry for micro-component*. PhD Thesis, 2008.

[16] G. S. Murugan, J. S. Wilkinson, and M. N. Zervas. Selective excitation of whispering gallery modes in a novel bottle microresonator. *Optics Express*, 17(14):11916–11925, Jul 2009. https://doi.org/10.1364/OE.17.011916. URL https://opg.optica.org/oe/abstract.cfm?URI=oe-17-14-11916.

[17] K. Ott, S. Garcia, R. Kohlhaas, K. Schüppert, P. Rosenbusch, R. Long, and J. Reichel. Millimeter-long fiber Fabry–Perot cavities. *Optics Express*, 24(9):9839–9853, May 2016. https://doi.org/10.1364/OE.24.009839. URL https://opg.optica.org/oe/abstract.cfm?URI=oe-24-9-9839.

[18] M. Pöllinger, D. O'Shea, F. Warken, and A. Rauschenbeutel. Ultrahigh-Q tunable whispering-gallery-mode microresonator. *Physical Review Letters*, 103:053901, Jul 2009. https://doi.org/10.1103/PhysRevLett.103.053901. URL https://link.aps.org/doi/10.1103/PhysRevLett.103.053901.

[19] M. Pöllinger. *Bottle microresonators for applications in quantum optics and all-optical signal processing.* PhD thesis, Mainz, Univ., Diss., 2011, 2010.

[20] S. Ravets, J. E. Hoffman, P. R. Kordell, J. D. Wong-Campos, S. L. Rolston, and L. A. Orozco. Intermodal energy transfer in a tapered optical fiber: optimizing transmission. *Journal of the Optical Society of America A*, 30(11):2361–2371, Nov 2013. https://doi.org/10.1364/JOSAA.30.002361. URL https://opg.optica.org/josaa/abstract.cfm?URI=josaa-30-11-2361.

[21] P. Romagnoli, M. Maeda, J. M. Ward, V. G. Truong, and S. Nic Chormaic. Fabrication of optical nanofibre-based cavities using focussed ion-beam milling: a review. *Applied Physics B*, 126(6):111, May 2020. ISSN 1432-0649. https://doi.org/10.1007/s00340-020-07456-x.

[22] C. Saavedra, D. Pandey, W. Alt, H. Pfeifer, and D. Meschede. Tunable fiber Fabry–Perot cavities with high passive stability. *Optics Express*, 29(2):974–982, Jan 2021. https://doi.org/10.1364/OE.412273. URL https://opg.optica.org/oe/abstract.cfm?URI=oe-29-2-974.

[23] C. Saavedra, D. Pandey, W. Alt, D. Meschede, and H. Pfeifer. Spectroscopic gas sensor based on a fiber Fabry–Perot cavity. *Physical Review Applied*, 18:044039, Oct 2022. https://doi.org/10.1103/PhysRevApplied.18.044039. URL https://link.aps.org/doi/10.1103/PhysRevApplied.18.044039.

[24] W. A. Snyder, J. D. Love, and et al.. *Optical waveguide theory* volume 175. Chapman and hall, London, 1983.

[25] D. Hunger, T. Steinmetz, Y. Colombe, C. Deutsch, T. W. Hänsch, and J. Reichel. A fiber Fabry–Perot cavity with high finesse. *New Journal of Physics*, 12(11):065038, 06 2010. https://doi.org/10.1088/1367-2630/12/6/065038.

[26] T. Steinmetz, Y. Colombe, D. Hunger, T. W. Hänsch, A. Balocchi, R. J. Warburton, and J. Reichel. Stable fiber-based Fabry–Pérot cavity. *Applied Physics Letters*, 89(11):111110, 09 2006. ISSN 0003-6951. https://doi.org/10.1063/1.2347892.

[27] M. Sumetsky. Optical bottle microresonators. *Progress in Quantum Electronics*, 64:1–30, 2019. ISSN 0079-6727. https://doi.org/10.1016/j.pquantelec.2019.04.001. URL https://www.sciencedirect.com/science/article/pii/S0079672719300072.

[28] M. Uphoff, M. Brekenfeld, G. Rempe, and S. Ritter. Frequency splitting of polarization eigenmodes in microscopic Fabry–Perot cavities. *New Journal of Physics*, 17(1):013053, Jan 2015. https://doi.org/10.1088/1367-2630/17/1/013053. URL https://dx.doi.org/10.1088/1367-2630/17/1/013053.

[29] C. Wuttke, M. Becker, S. Brückner, M. Rothhardt, and A. Rauschenbeutel. Nanofiber Fabry–Perot microresonator for nonlinear optics and cavity quantum electrodynamics. *Optics Letters*, 37(11):1949–1951, Jun 2012. https://doi.org/10.1364/OL.37.001949. URL https://opg.optica.org/ol/abstract.cfm?URI=ol-37-11-1949.

[30] C. Wuttke. *Thermal excitations of optical nanofibers measured with a fiber-integrated Fabry–Pérot cavity.* PhD thesis, Mainz, Univ., Diss., 2014, 2014.

4 Cavity-QED

The role of light or optical photons is of utmost importance in communication and quantum information sciences [14] due to several fascinating facts:

– *Mobility:* Photons possess inherent mobility at the speed of light, allowing for easy and high-speed transportation through optical fibers or in free space. This makes them a natural and convenient choice as information carriers for both classical and quantum communication.

– *Experimental simplicity:* Compared to other systems like atomic, semiconductor, and superconducting systems, the generation, manipulation, and detection of quantum states using photons are relatively more straightforward.

– *Negligible interaction:* Photons exhibit minimal interaction with their surroundings, enabling the creation of clean and decoherence-free photonic qubits. Additionally, information can be encoded effortlessly in various degrees of freedom, such as polarization, frequency, temporal, and spatial modes.

– *Decoherence resilience:* Photons at optical wavelengths typically possess energies in the range of a few electron volts. This characteristic grants them a significant advantage in terms of high contrast compared to background thermal noise at typical ambient temperatures. Consequently, photons can be transported with minimal susceptibility to environmental conditions and do not experience significant decoherence.

– *Compatibility with existing infrastructure:* The utilization of a photon-based interface for quantum technology offers the advantage of aligning with the already established optical fiber technology employed in telecommunications. This compatibility enables the seamless integration of quantum systems with the existing infrastructure, allowing for efficient scaling and practical implementation in real-world applications.

Although photons are considered one of the best and excellent candidates for information transfer, they still face challenges when used as carriers of quantum information for long-distance quantum information processing. These challenges arise from attenuation caused by absorption and scattering during propagation. Traditional methods of light amplification, used in classical communication, are unsuitable for preserving quantum states due to the impossibility of cloning them, which prohibits classical signal amplification as a solution to overcome signal attenuation. In order to address this issue, compensating strategies based on quantum repeaters are necessary [5]. Quantum repeaters, discussed in Chapter 5, allow for enhancing entanglement distribution over longer distances [7]. Quantum memories play a critical role as fundamental components for implementing quantum repeater strategies. They involve manipulating photon wave packets by delaying, storing, or shaping them, relying on the interaction between light and matter [15]. Furthermore, in the context of quantum computing, achieving universal quantum gates poses a challenge for photons due to their lack of inherent interaction

https://doi.org/10.1515/9783110636260-004

with each other. However, by leveraging light-matter interactions, it becomes possible to enable nonlinear interactions among photons. As a result, achieving controlled interactions between light and matter becomes one of the most crucial tasks in quantum information processing.

Cavity quantum electrodynamics (Cavity-QED) offers a foundational framework for exploring the intricate interaction between light and matter. It provides a powerful tool for studying and controlling the interaction of photons with matter-based systems. This chapter systematically introduces the concepts of cavity-QED physics and its application in realizing quantum memories and single-photon sources.

4.1 Cavity quantum electrodynamics

Cavity quantum electrodynamics (Cavity-QED) is a powerful framework that examines the dynamics of the interaction between a confined electromagnetic field and atoms or particles within a resonator (Figure 4.1). This interaction takes place under conditions where the quantum nature of photons becomes remarkably significant. By considering dissipation channels associated with both resonators and atoms, cavity-QED offers a versatile platform for exploring fundamental aspects of quantum mechanics in open quantum systems. It facilitates the engineering of quantum states through coherent interactions between light and matter, as well as the study of measurement-induced coherence dynamics. The remarkable contributions of cavity-QED to the field of quantum sciences were acknowledged with the awarding of the 2012 Nobel Prize to Serge Haroche [11].

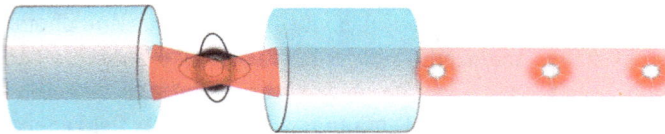

Figure 4.1: A Fabry–Perot cavity used for interfacing photonic quantum information into the atomic states using cavity-QED physics.

An optical resonator plays a vital role in a cavity-QED system. We have discussed in Chapter 2 about optical resonators, where we explored the possibility of light confinement using optical cavity. In the subsequent section, we will provide a brief reintroduction to the key parameters of a cavity and present a quantum description of light within a resonator.

4.1.1 Field quantization

In the context of optical resonators, the confinement of light is achieved through the interference of light beams that undergo multiple reflections within the resonator. The characteristics of a mirror-based optical cavity are determined by the quality of the mirror surfaces and the distance between them. The reflectivity of the mirror coatings plays a crucial role in defining the cavity properties. Two important parameters that quantify the resonator properties are finesse and cavity linewidth.

Finesse represents the number of times a photon bounces between the cavity mirrors before being lost, either by leaking out through the mirrors or being absorbed by the mirror coatings. It can be calculated as

$$F = \frac{2\pi \sqrt{R}}{1 - R},$$

(4.1)

where R is the reflectivity of the mirror coating. On the other hand, the cavity linewidth indicates the rate at which photons can leak out of the cavity in a steady-state condition.

For a Fabry–Perot cavity with mirror separation l and equal reflectivities for both mirrors, the free spectral range (FSR), which represents the separation between adjacent resonant frequencies of the cavity, is given by

$$\nu_{\text{FSR}} = \frac{c}{2l},$$

(4.2)

where c is the speed of light. The linewidth of the cavity can be related to the finesse and FSR as $2\kappa = \frac{\nu_{\text{FSR}}}{F}$. The transmission profile near the cavity resonance follows a Lorentzian shape with a linewidth of 2κ (note, notation $\Delta\nu_{FWHM}$ is used in Chapter 2) which provides information about the temporal dynamics of the cavity field, specifically the rate of decay of the cavity field.

In the regime of low photon numbers, the quantum mechanical description becomes important for the light field inside an optical cavity. In this regime, the photon field exhibits discrete energy states. Analogous to the energy levels of a quantum harmonic oscillator, the photon number states can be represented by the state vector $|n\rangle$, where n represents the number of photons. The Hamiltonian for the cavity field energy can be written as (see Appendix C):

$$H_{\text{cav}} = \hbar\omega_{\text{cav}}\left(\hat{a}^{\dagger}\hat{a} + \frac{1}{2}\right),$$

(4.3)

where $\hat{a}^{\dagger}$ and $\hat{a}$ are the creation and annihilation operators for a photon inside the cavity. These operators also contain information about the spatial mode profile of the cavity field, such as Laguerre–Gaussian or Hermite–Gaussian modes, commonly used in Fabry–Perot cavities with curved mirrors.

4.1.2 Jaynes–Cumming–Hamiltonian

Cavity-QED deals with the interaction mechanism between the light and a quantum particle inside an optical resonator. The Jaynes–Cummings model, developed in 1963,

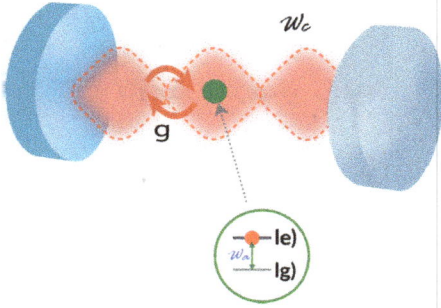

Figure 4.2: A two-level atomic system is coupled to the mode of a cavity.

describes the interaction of a two-level atom with a single mode of quantized electromagnetic radiation field. As illustrated in Figure 4.2, when a two-level atom is placed inside an optical resonator, it can exchange energy coherently with the cavity field. This interaction allows the atom to absorb a photon from the cavity field and transition to an excited state or emit a photon into the cavity and transition to the ground state. In an ideal scenario of negligible decay rates, such as no atomic spontaneous emission or leakage through the cavity, the emitter-cavity interaction is the strongest and coherent. In such a dissipation-less system, the coherent exchange of energy between the two systems is described by the JC–Hamiltonian. There are three parts of the Hamiltonian:

- *The cavity field:* The quantization of the electromagnetic field inside the cavity gives rise to cavity eigenstates that resemble a harmonic oscillator. The Hamiltonian for the cavity field is given by

$$H_c = \hbar\omega_c \hat{a}^\dagger \hat{a}, \tag{4.4}$$

where $\hat{a}^\dagger$ and $\hat{a}$ are the creation and annihilation operators for a photon inside the cavity, and ω_c is the frequency of the cavity mode.

- *Two-level atom:* The Hamiltonian for a two-level atom can be expressed as

$$H_a = \hbar\omega_a \frac{\hat{\sigma}_z}{2} \tag{4.5}$$

where $\hbar\omega_a = \hbar(\omega_e - \omega_g)$. $\hbar\omega_e$ and $\hbar\omega_g$ are the energies of the two atomic eigenstates. The form of the above Hamiltonian can be briefly explained as follows. Consider a two-level system with eigenenergies $\hbar\omega_g$ and $\hbar\omega_e$ for the ground and the excited state $|g\rangle$ and $|e\rangle$, respectively. The Hamiltonian for the atom in the rest frame is

$$H_a = \hbar\omega_e |e\rangle \langle e| + \hbar\omega_g |g\rangle \langle g| \tag{4.6}$$

The state vector space for a two-level system has four linearly independent operators, which can be chosen in terms of standard spin operators

$$\sigma_0 = |e\rangle \langle e| + |g\rangle \langle g| \rightarrow \begin{pmatrix} 1 & 0 \\ 0 & 1 \end{pmatrix}$$

$$\sigma_z = |e\rangle \langle e| - |g\rangle \langle g| \rightarrow \begin{pmatrix} 1 & 0 \\ 0 & -1 \end{pmatrix}$$

$$\sigma^+ = |e\rangle \langle g| \rightarrow \begin{pmatrix} 0 & 1 \\ 0 & 0 \end{pmatrix} \tag{4.7}$$

$$\sigma^- = |g\rangle \langle e| \rightarrow \begin{pmatrix} 0 & 0 \\ 1 & 0 \end{pmatrix}$$

With reorganizing equation (4.6), one can get the following form in terms of Pauli operators:

$$H_a = \hbar \omega_a \frac{\hat{\sigma}_z}{2} + \hbar(\omega_e + \omega_g)\sigma_0 \tag{4.8}$$

Finally, energy rescaling with an offset of $\frac{1}{2}(\omega_e + \omega_g)$, we obtain the atomic Hamiltonian given in equation (4.5).

- *Interaction term:*

$$H_{\text{int}} = \hbar g(\sigma_{eg} a + a^\dagger \sigma_{ge}) \tag{4.9}$$

where $\sigma_{eg} = \sigma^+$ and $\sigma_{ge} = \sigma^-$ are the atomic raising and lowering operators. $g = \psi_c(r) g_0$, is the atomic cavity interaction strength, which includes terms from the quantization of the electric field and the atomic dipole moment operator

$$g_0 = \left(\sqrt{\frac{\omega}{2\epsilon_0 V}} \right) d, \tag{4.10}$$

and $\psi_c(r)$ represents the spatial profile of the cavity mode field. V is the mode volume of the cavity and d is the dipole moment for the atomic transition (see Appendix C).

The interaction term H_{int} captures the processes of energy exchange between the atom and the cavity field. The first term in the parenthesis ($\sigma_{eg}a$) represents the annihilation of a photon from the cavity field (a) and the excitation of the atom from the ground state to the excited state (σ_{eg}). The second term ($a^\dagger \sigma_{ge}$) corresponds to the creation of a photon in the cavity field ($a^\dagger$) and the deexcitation of the atom from the excited state to the ground state (σ_{ge}).

In a closed system without any dissipation channels, the atom and the cavity can undergo coherent oscillations between their respective states. The interaction term allows for the exchange of energy between the cavity and the atomic states. By representing the states of the electromagnetic field and the atom as product states of their individual Hilbert spaces (e. g., $|g, n\rangle$ for the ground state of the atom and n photons in the cavity, and $|e, n-1\rangle$ for the excited state of the atom and $n-1$ photons in the cavity), one can describe the dynamics of the system in terms of these states:

$$|g, n\rangle = |g\rangle \otimes |n\rangle, \quad |e, n-1\rangle = |e\rangle \otimes |n-1\rangle \tag{4.11}$$

For a given excitation number $n \geq 1$ inside the cavity field, there is a coherent energy transfer between the atom and the cavity field. In the collective Hilbert space, cavity couples $|g, n\rangle$ and $|e, n-1\rangle$ via the total Hamiltonian

$$H = H_a + H_c + H_{\text{int}} \tag{4.12}$$

In rotating wave approximation, the Hamiltonian coupling of the two states $|g, n\rangle$ and $|e, n-1\rangle$ for a laser with angular frequency ω is

$$H_n \begin{pmatrix} |g, n\rangle \\ |e, n-1\rangle \end{pmatrix} = \hbar \begin{pmatrix} n\,\omega_{\text{cav}} - \frac{1}{2}(\omega_a) & g\sqrt{n} \\ g\sqrt{n} & (n-1)\omega_{\text{cav}} + \frac{1}{2}(\omega_a) \end{pmatrix} \begin{pmatrix} |g, n\rangle \\ |e, n-1\rangle \end{pmatrix} \tag{4.13}$$

Solving the eigenvalue problem of this Hamiltonian yields the following eigenfrequencies:

$$\omega_n^{\pm} = \omega_{\text{cav}}\left(n + \frac{1}{2}\right) + \frac{1}{2}\left(\Delta_{\text{cav}} \pm \sqrt{4ng_0{}^2 + \Delta_{\text{cav}}{}^2}\right) \tag{4.14}$$

Here, $\Delta_{\text{cav}} = \omega_e - \omega_g - \omega_{\text{cav}}$ is the detuning between the cavity and atom within each n-fold excitation Hilbert space. The frequency splitting between two eigenstates for a given photon number from equation (4.14) is

$$\Omega_{\text{cav}} = \sqrt{4ng_0{}^2 + \Delta_{\text{cav}}{}^2} \tag{4.15}$$

This is the effective Rabi frequency with which the population oscillates between the atomic state and the cavity excitation field via states $|g, n\rangle$ and $|e, n-1\rangle$. For photon number state $n = 1$, the coherent oscillation is between states $|g, 1\rangle$ and $|e, 0\rangle$, which means an atom in excited states with a vacuum state in the cavity is enough to start the oscillation and is therefore called "Vacuum-Rabi oscillation (VRS). The VRS frequency is $\Omega_{\text{cav}} = \sqrt{4g_0{}^2 + \Delta_{\text{cav}}{}^2}$. It is important to note that at zero detuning between the atom and the cavity $\Delta_{\text{cav}} = 0$, the atom-cavity coupling strength decides the rate of the excitation oscillation $\Omega_{\text{cav}} = 2g_0$. Therefore, for strong atom-cavity coupling, g_0, the photon number states are split into nondegenerate dressed states. The eigenstates of the Hamiltonian are as follows:

$$|n, +\rangle = \cos\frac{\alpha_n}{2}\,|g, n\rangle + \sin\frac{\alpha_n}{2}\,|e, n-1\rangle \tag{4.16}$$

$$|n, -\rangle = \sin\frac{\alpha_n}{2}\,|g, n\rangle - \cos\frac{\alpha_n}{2}\,|e, n-1\rangle \tag{4.17}$$

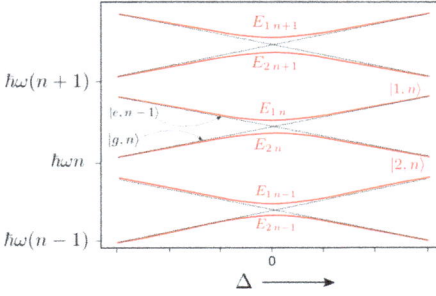

Figure 4.3: Dressed energy states of an atom inside a cavity.

where $a_n = \tan^{-1}(\frac{\Omega\sqrt{n}}{\Delta})$ and $\Delta = \omega_a - \omega$. As shown in Figure 4.3, each manifold of a given photon number n is split into two dressed states.

4.2 Cavity-QED regimes

In the previous section, we discussed the interaction between an atom and the quantized radiation field in a closed quantum system. This interaction results in a coherent exchange of energy between the cavity field and the atomic excitation, leading to vacuum Rabi oscillations. The oscillation frequency, denoted as g, represents the strength of the atom-cavity coupling.

However, in real systems, there are losses associated with both the atomic excitation and the photon field excitation. The atomic excitation can decay with a rate Γ through spontaneous emission outside the cavity modes. Similarly, the light field inside the cavity leaks out with a rate κ, which depends on the properties of the cavity mirrors.

In cavity, quantum electrodynamics (Cavity-QED), the dynamics of the system can be categorized into three regimes based on the relative magnitudes of the atomic decay rate, cavity decay rate, and atom-cavity coupling rate. To capture these different rates in a single parameter, we use the concept of cooperativity. The cooperativity parameter, denoted as C, is defined as

$$C = \frac{4g^2}{\Gamma\kappa} \tag{4.18}$$

- *Strong coupling regime:* When the cooperativity is greater than 1, i.e., ($C > 1$) and $g > (\kappa, \Gamma)$, the system operates in the strong coupling regime (Figure 4.4). In this regime, the system dynamics exhibit coherent vacuum Rabi oscillation because the reversible atom-photon coupling rate g is much more than the rate of the irreversible decay rates (κ, Γ). Due to the coherent energy exchange between the atom and the cavity in this regime, the atom-cavity dynamic is very close to the formalism described by the Jaynes–Cummings–Hamiltonian, and vacuum Rabi oscillation can be experimentally observed.

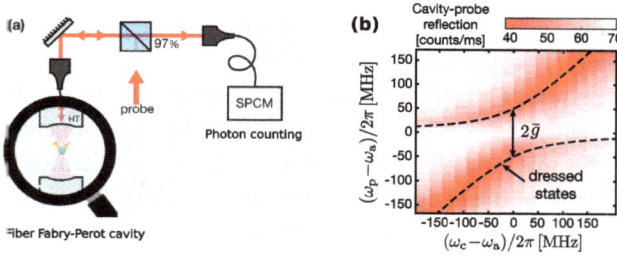

Figure 4.4: An atom-cavity system in the strong coupling regime is reported in reference [9]. Figure (a) depicts the setup used for measuring the response of the atom-cavity using a probe laser beam. The reflected light from the cavity is measured on a single-photon-counting module (SPCM). Figure (b) demonstrates the change in cavity reflection with respect to both probe detuning ($\omega_p - \omega_a$) and cavity detuning ($\omega_c - \omega_a$). An avoided crossing or vacuum-Rabi-splitting (VRS) is observed as a typical signature of strong coupling. The coupling rate g (or VRS) at $\omega_c = \omega_a$ measured in the experiment results in a cooperativity parameter $C = 7.2$. This figure is adapted with permission from [9].

– *Purcell regime:* If $C > 1$ and $\kappa > g > \Gamma$, the leakage of a photon through the cavity is faster than the atom-cavity coupling. Therefore, vacuum Rabi oscillation does not occur as the photon emitted into the cavity mode leaks out of the cavity before reabsorption in the atom. However, the emission properties of the atom are strongly modified due to changes in the local electric field density. The rate of emission can be enhanced or suppressed depending on the detuning of the cavity from the atomic transitions. The rate of the single-photon emission in this regime is [24]:

$$\Gamma_c = \frac{g^2 \kappa}{\kappa^2 + \Delta_c^2},$$ (4.19)

$\Delta_c = \omega_a - \omega_c$ is the detuning of the cavity resonance ω_c from the atomic transition ω_a frequency. When $\Delta_c = 0$, the relative emission rate into the cavity compared to free space is double the cooperativity parameter:

$$\frac{\Gamma_c}{\Gamma} = 2C$$ (4.20)

– *Weak coupling regime* When $C < 1$, either the cavity or the spontaneous atomic emission dominates, and the incoherent decay prohibits reversible energy transfer between the atom and the cavity.

4.2.1 Purcell effect

The emission characteristics of a luminescent material can undergo significant changes when it is enclosed within a cavity. This phenomenon, known as the Purcell effect, was first proposed by Edward Purcell in 1946 in relation to the enhanced spontaneous emission rate of a nuclear spin coupled to a resonant electric circuit at radio frequencies

[23]. The Purcell effect occurs due to the cavity's ability to modify the local electromagnetic field density of states surrounding the emitter. This alteration influences the number of electromagnetic modes per unit frequency and per unit volume available for the emitter to radiate into. Consequently, the emission properties arise from the interaction between the emitter and the effective local electromagnetic density of states, which is strongly influenced by the strength of the emitter-cavity coupling. In certain cases involving strong atom-cavity interaction, the emission can be completely suppressed or greatly enhanced.

To quantitatively evaluate the modification of the excited decay rate of an atom within a cavity, second-order perturbation theory and the Fermi golden rule can be employed to calculate the local density of the electromagnetic field. The modified emission rate of an emitter in the presence of an optical resonator can be expressed as

$$\Gamma_g = \frac{2\pi\mu_{12}{}^2 E_0{}^2}{\hbar}\rho(\omega_0) \tag{4.21}$$

where μ_{12} is the dipole strength of the optical transition, E_0 is the electric field amplitude, and $\rho(\omega_0)$ is the electromagnetic local density of states at frequency ω_0. $\rho(\omega_0)$ quantifies the number of electromagnetic modes per unit frequency and per unit volume available for the emitter for radiation emission. The multipass effect due to high finesse and focusing of the light at the waist of a resonator can significantly modify available states $\rho(\omega_0)$.

The ratio of the modified emission rate to the emission rate in free space became known as the Purcell factor. When an emitter is coupled to the maximum intensity of the field, with its dipole aligned with the polarization of the cavity mode, the Purcell factor is typically given by

$$F_P = \frac{\Gamma_c}{\Gamma_0} = \frac{3Q\lambda^3}{4\pi^2 V_0} \tag{4.22}$$

where Q is the quality factor of the resonator, and V_0 is the mode volume of the resonator.

For a single emitter coupled to a cavity with a coupling strength of g, the cavity decay rate of κ, and emitter spontaneous emission rate of γ ($\gamma = \frac{\Gamma}{2}$), the Purcell factor is twice the cooperativity parameter:

$$F_p = 2C = \frac{g^2}{\kappa\gamma} \tag{4.23}$$

For cavity-enhanced single-photon emission, it is important to maximize the Purcell factor. The rate of the photon emission into the cavity mode is given by

$$P_{\text{emission}} = \frac{F_p}{F_p + 1} \tag{4.24}$$

Interaction of a single emitter within a cavity, including the decay terms both from the spontaneous emission of atom (excited state $|e\rangle$) as well as the cavity decay, can be described by introducing the non-Hermitian decay terms into the Jaynes–Cumming–Hamiltonian [9]:

$$H = H_{JC} - i\hbar\gamma\,|e\rangle\,\langle e| - i\hbar\kappa\hat{a}^\dagger\hat{a} \qquad (4.25)$$

The Purcell effect has important implications in the realization of single-photon sources because this effect allows the photon emission rate into a well-defined cavity mode (Figure 4.5) by engineering the properties of the resonator. Also, quantum emitters with complex energy level structures benefit from enhancing a specific desired transition while simultaneously suppressing the others.

(a) **(b)**

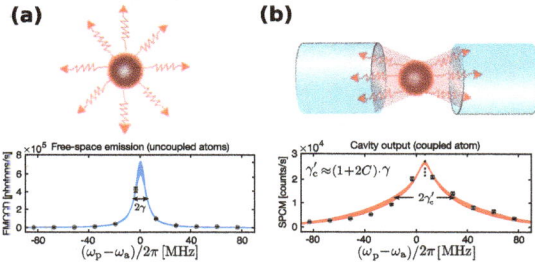

Figure 4.5: Modification of the emission rate and the directionality of the photon from a quantum emitter inside a cavity due to the Purcell effect. The plot in Figure (a) shows the linewidth of a single atom in free space ($\propto \gamma$) while the plot in Figure (b) shows the modified linewidth ($\propto \gamma_c' = (1 + 2C)\gamma$), where C is the cooperativity parameter and, ω_p and ω_a are the frequencies of the probe laser and the atomic transition respectively. The plots in this figure are adapted from reference [9] with permission from the American Physical Society.

4.3 Three-level system in a cavity

Atomic systems with three-level configurations, typically with two quasi-stable ground states coupled independently with an excited state, are the most frequently used configuration in quantum experiments. Termed as λ-system, such a three-level system can be used to realize the quantum interference effects when atomic excitation via two different transition paths is implemented. Here, we describe a λ-system configuration used in the cavity-QED for quantum information processing.

An atom with λ-configuration level scheme, as shown in Figure 4.6, has two ground states $|g_1\rangle$, $|g_2\rangle$, and an excited state $|e\rangle$. The transition frequency between the ground states and the excited states are

$$\omega_{eg_1} = \omega_e - \omega_{g_1} \quad \text{and} \quad \omega_{eg_2} = \omega_e - \omega_{g_2} \qquad (4.26)$$

Transition ω_{eg_1} is driven by a classical light field frequency ω_L and Rabi frequency Ω. One branch of the λ transition, i. e., $|g_2\rangle \rightarrow |e\rangle$, is driven by the cavity field of frequency ω_{Cav}.

(a)

(b)

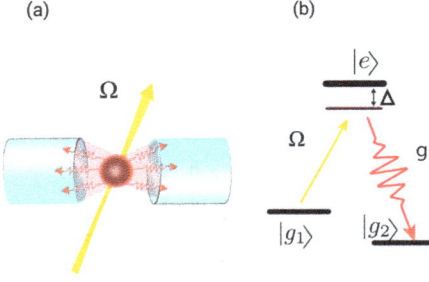

Figure 4.6: Figures (a) and (b) shows a three level λ-system coupled to a cavity. The atom-cavity coupling strength along one of the transitions is g and the other transition is coupled via a laser (control laser) traversing the atom through the side of the cavity. Ω represents the Rabi frequency of the control laser while Δ represents the detuning from the excited state.

The detuning or the difference with respect to the exact atomic transition frequency is defined as follows:

$$\Delta_L = \omega_{eg_1} - \omega_L \tag{4.27}$$

$$\Delta_{\text{cav}} = \omega_{eg_2} - \omega_{\text{cav}} \tag{4.28}$$

Assuming no cross-coupling between the transitions, one can write the interaction Hamiltonian as follows:

$$H_{\text{int}} = \hbar \left[\Delta_L \, |g_1\rangle \, \langle g_1| + \Delta_{\text{cav}} \, |g_2\rangle \, \langle g_2| - \frac{\Omega}{2} (|e\rangle \, \langle g_1| + |g_1\rangle \, \langle e|) - g_0 (|e\rangle \, \langle g_2| \, a + \hat{a} \, |g_2\rangle \, \langle e|) \right] \tag{4.29}$$

For any given excitation number n, one can deduce that this Hamiltonian couples only three states $|g_1, n-1\rangle$, $|e, n-1\rangle$, $|g_2, n\rangle$. For a Raman resonant configuration where

$$\Delta_L = \Delta_{\text{cav}} = \Delta, \tag{4.30}$$

there exist the eigenfrequencies of the triplet state are as follows:

$$\omega_n^{\,0} = \omega_{\text{Cav}} \left(n + \frac{1}{2} \right) \tag{4.31}$$

and

$$\omega_n^{\pm} = \omega_{\text{cav}} \left(n + \frac{1}{2} \right) + \frac{1}{2} (\Delta \pm \sqrt{4n g_0^2 + \Omega^2 + \Delta^2}) \tag{4.32}$$

These three triplet eigenstates are defined as follows:

$$|\phi_n^{\,0}\rangle = \cos \Theta \, |g_1, n-1\rangle - \sin \Theta \, |g_2, n\rangle \tag{4.33}$$

$$|\phi_n^{\,+}\rangle = \cos \phi \sin \Theta \, |g_1, n-1\rangle - \sin \phi \, |e, n-1\rangle + \cos \phi \cos \Theta \, |g_2, n\rangle \tag{4.34}$$

$$|\phi_n^{\,-}\rangle = \sin \phi \sin \Theta \, |g_1, n-1\rangle + \cos \phi \, |e, n-1\rangle + \sin \phi \cos \Theta \, |g_2, n\rangle \,. \tag{4.35}$$

Here, the mixing angle Θ and ϕ are given as follows:

$$\tan\Theta = \frac{\Omega}{2g_0\sqrt{n}}, \quad \tan\phi = \frac{\sqrt{4ng_0{}^2 + \Omega^2}}{4n} \tag{4.36}$$

$|\phi_n{}^0\rangle$ does not have any contribution from the excited state and, therefore, it does not decay via spontaneous emission. Therefore, this state is also termed a dark state.

4.3.1 Stimulated Raman adiabatic passage

Stimulated Raman adiabatic passage (STIRAP) is a technique used in quantum optics to transfer the population between different quantum states or levels of a quantum system. It is a coherent and efficient method for manipulating quantum states, particularly in multilevel atomic or molecular systems.

In the simplest implementation of this process, two laser pulses, often referred to as the pump and Stokes fields with optimized pulse shape and sequence, are used to facilitate the population transfer between the two discrete quantum states via an intermediate state. It allows efficient and selective transfer of atomic population between states while protecting the quantum coherence against spontaneous emission. As shown in Figure 4.7, a three-level system with two quasi-stable ground states can be connected via an excited state. In normal circumstances, each transition $g_{1,2} \to e$ can incoherently decay to the two ground states via spontaneous emission. However, in an ideal STIRAP process, the quantum interference between the different possible paths allows the complete transfer will negligible effect of spontaneous emission.

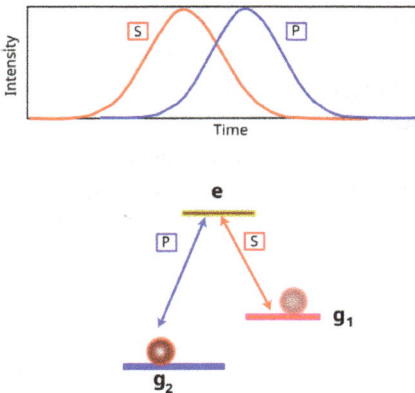

Figure 4.7: STIRAP process allows the transfer of atomic population between two ground states in the picture. The intensity of Stokes and pump laser pulses follows adiabatic condition and arrive with the temporal profile as shown in the figure.

Considering the two laser radiation fields connecting the two ground states to the excited states have a Rabi frequency, Ω_S for $|g_1\rangle \to |e\rangle$, and Ω_P for $|g_2\rangle \to |e\rangle$ respectively. As shown in Figure 4.7, the atomic population initially is at $|g_1\rangle$. However, the laser labeled as Stokes arrives first. The S-pulse adiabatically rises. Close to the maximum intensity of *S*-pulse, another laser field labeled as pump start rising. *S*-pulse starts adiabatically going down. *P*-pulse follows the same pulse shape with a delayed time [28]. Population transfer with very high efficiency is possible with this method. The key principle behind STIRAP is adiabaticity. Adiabaticity ensures that the population transfer occurs without excitations or losses to other states, even in the presence of experimental imperfections or fluctuations. By slowly varying the laser intensities and frequencies, the system remains in its dark state throughout the transfer, enabling efficient and robust population transfer.

Once the system is prepared in the dark state, the Stokes field is applied. The Stokes laser is tuned to couple the intermediate state $|2\rangle$ to the final state $|3\rangle$. This coupling is achieved by matching the energy difference between $|2\rangle$ and $|3\rangle$ with the frequency of the Stokes laser. The application of the Stokes field induces a coherent transfer of population from $|2\rangle$ to $|3\rangle$, completing the desired population transfer. We are interested in a STIRAP process, which is used for a three-level system in a cavity-QED setting. This process, known as vacuum STIRAP, is an important technique for single-photon generation and storage using a cavity coupled to an atom. We have already seen in Section 4.3, how a three-level λ-system with one of the transitions driven by a cavity field and the other with a classical laser field, has modified eigenstates $|\phi_n^{\,0}\rangle$, $|\phi_n^{\,-}\rangle$, and $|\phi_n^{\,+}\rangle$. These three dressed eigenstates are involved in adiabatically transitioning from one atomic ground state to the other using STIRAP, resulting in the generation of a single photon in the cavity mode.

4.3.2 Single-photon sources

As the name suggests, a single-photon state is the excitation of an electromagnetic field for which the photon number statistics have a mean value of one with a null variance. This means a single-photon state produces always a single event corresponding to one photon incident on an ideal photon number resolving detector. The variance of photon number distribution and photon number statistics differentiates a single-photon state from conventional light sources, such as an incandescent lamp with a thermal state and a laser with a coherent state. To differentiate the three different light sources, one can note the following:

–   Thermal light state exhibits the following photon number distribution at a given frequency, also known as Bose–Einstein distribution,

$$P(n) = \frac{1}{\bar{n}+1}\left(\frac{\bar{n}}{\bar{n}+1}\right)^n. \tag{4.37}$$

Variance is given by

$$\Delta n^2 = \sum_{n=0}^{\infty}(n-\bar{n})^2 P(n) = \bar{n} + \bar{n}^2 \tag{4.38}$$

- Coherent light or laser light has inherently Poissonian photon statistics. It means for a laser beam with perfectly constant power and with an average photon number $\bar{n}$, the photon number distribution is given by

$$P(n) = \frac{\bar{n}^n}{n!} e^{-\bar{n}} \tag{4.39}$$

Variance is given by

$$\Delta n^2 = \sum_{n=0}^{\infty} (n - \bar{n})^2 P(n) = \bar{n} \tag{4.40}$$

- Sub-Poissonian light fields are the nonclassical light fields and are characterized by

$$\Delta n^2 < \bar{n} \tag{4.41}$$

A single-photon state is a photon number state with $\Delta n = 0$.

A single-photon source in its idealistic realization enables user-defined temporal profile and deterministic on-demand generation of single photons with 100 % probability and null probability for multiphoton emission. The generation rates can be arbitrarily large with indistinguishable subsequent emitted photons. The source should be photostable for all time scales, which means its emission pattern does not have any bleaching or blinking effect.

Practically single-photon sources, which can be used in quantum information processing, are constrained in different characteristics. These constrains depend on the method and the particular system used. However, certain crucial parameters can be identified and optimized based on their intended application.

4.3.3 Probabilistic vs. deterministic single-photon sources

Single-photon sources generated by the spontaneous parametric down-conversion (SPDC) method are known as *probabilistic single-photon sources*. This is because of the nature of the nonlinear process, which probabilistically generates a photon pair in two different spatial modes. The process also generates a vacuum state with no photons and sometime multi-photon pairs. Therefore, heralding is used to have a quasi-deterministic single photon as follows. Using a heralding detector to detect one of the photons from the pair allows for determining the incidents when photon pairs are generated, and the simultaneous usage/detection of the other photon of the pair represents a single-photon state. SPDC single-photon sources are the least complex system without a complicated cooling, vacuum, or cryogenic system. The SPDC requires excitation of nonlinear optical processes that convert high energy pump photons into a pair of photons (signal and

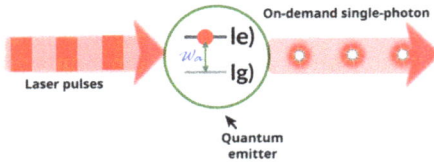

Figure 4.8: Excitation of a single-quantum emitter and subsequent relaxation process leads to a single-photon state.

idler) in an SPDC crystal, e. g., Beta-Barium-Borate (BBO), Potassium Titanyl Phosphate (KTP), Periodically Poled Lithium Niobate (LN), and many other systems.

Deterministic single-photon sources is expected to have exactly one photon deterministically emitted following an excitation process as shown in Figure 4.8. Single atoms and ions, color centers in diamonds, and quantum dots are some of the systems, which can generate deterministic single photons.

Emission from a single-quantum emitter is a deterministic single-photon generation process because a single emitter can have only one excitation, which is released as a single photon through relaxation. The efficiency in such systems can be very high and only limited by collection efficiency and losses in the outcoupling process. Both of these can be significantly enhanced by using optical cavities surrounding the emitter.

Optical cavities are one of the crucial tools to direct, in some cases also enhance, the photon emission into a well-defined cavity mode. Therefore, much research is ongoing on developing cavity-integrated systems, and fiber-based resonators are one of the promising platforms for that. In the context of this book, we are interested in the research endeavors following the approach of fiber-based cavity coupling for single-photon sources.

4.4 Fiber cavity-based single-photon sources

Fiber-based cavities offer several advantages, including a small mode volume, high finesse, and open optical access to the quantum emitters within the cavity. Direct fiber coupling provides a convenient way to interface the light with the emitter [9]. In the context of this book, we would like to discuss some of the quantum emitters, which have been successfully interfaced to a fiber cavity for single-photon generation.

4.4.1 Single-quantum emitter

A single-quantum emitter can emit only one photon at a time. This can include single atoms, ions, molecules, or any other emitter with two optically addressable states. Typically, one of the internal states may be a stable or a quasi-stable ground state and another state (excited state) with radiative decay. By preparing the emitter in the excited state through an externally controlled excitation process, the subsequent decay from the excited state can be channeled into a cavity mode, generating high-quality single photons. The Purcell effect can be employed by selecting appropriate cavity parameters to enhance the emission rates and collection efficiency from the emitter into the well-defined mode [9, 18, 12]. The temporal shape of a single photon emitted from a two-level system is determined by the effective radiative lifetime of the emitter inside the cavity. In this regard, a three-level scheme provides much more control over the single-photon temporal profile described below.

4.4.2 Single atom/ion coupled to a fiber cavity

A three-level system with λ configuration is one of the most used systems for studying the quantum effects of light-matter interface. Figure 4.6 illustrates an example of this configuration, where two metastable ground states (g_1 and g_2) are coupled to an excited state (e). The λ configuration typically comprises these two ground states and an excited state. In the context of cavity quantum electrodynamics (QED), as discussed in the previous section, a three-level system plays a crucial role in generating single photons.

In this configuration, a three-level emitter is coupled to a cavity through one of the transitions in the λ configuration. The cavity resonance is set to the transition frequency $|g_1\rangle \rightarrow |e\rangle$. By adiabatically exciting the transition $|g_2\rangle \rightarrow |e\rangle$, a single photon is emitted into the $|e\rangle \rightarrow |g_1\rangle$ transition along the cavity mode. The presence of the cavity facilitates the efficient extraction of photons into a well-defined cavity mode. Moreover, the Purcell factor enhances the photon emission rate into the cavity mode, as explained in Section 4.2.1. The Purcell factor, denoted as F_P, is given by the equation:

$$F_P = \frac{2C}{2C + 1} \tag{4.42}$$

To achieve higher emission rates, it is desirable to have a strong coupling between the atom and the cavity. Although single-quantum emitters like single atoms provide single photons with high spectral purity, certain challenges are associated with their practical implementation. Controlling the atom's position at the cavity center often requires complex vacuum setups, laser cooling, and trapping techniques. Another drawback of deterministic single-photon sources is the absence of a heralding signal. Therefore, it is necessary to achieve efficiencies close to 100 % for practical applications.

However, an interesting experiment has used two orthogonal fiber cavities [4], which can generate a heralding signal as follows.

Two FFPCs with their axes orthogonal to each other and centers coinciding are used (see Section 4.4.4.1). A single-neutral rubidium atom is first laser-cooled and trapped, and then transferred to the center of the cavities. As the atom forms a λ-system simultaneously with both the cavities with strong coupling, the transfer of atomic population between ground states generates a herald signal for the other cavity. This scheme demonstrates an important milestone for cavity-based deterministic single-photon sources and quantum memories with signal heralding possibility.

4.4.2.1 Quantum dots coupled to a cavity

Qunatum-dots (Qdots) are semiconductor-based quantum emitters. They have a very small semiconductor region with a few tens of nanometer sizes, which is surrounded by another larger bandgap material that forms an interesting energy configuration. The potential barrier created by surrounding material confines electrons or holes in the conduction or the valence band of the Qdot. The resulting delta-function density of states for charge carriers provides isolated optical transitions, similar to atoms. For this reason, Qdots are sometimes also termed artificial atoms. Typically, the energy or the wavelength of the emitted photons from a Qdot depends upon the bandgap, size, and other controlled strain effects on the Qdot system.

One of the key advantages of quantum dot-based single-photon sources is their ability to emit photons at specific wavelengths. By engineering the size and composition of the quantum dot, the emission wavelength can be precisely controlled. This tunability allows for compatibility with various applications, such as quantum communication and quantum computing, which often require specific photon wavelengths. Moreover, quantum dots can exhibit excellent photon indistinguishability, a crucial property for many quantum information processing tasks [3, 26]. By carefully controlling the quantum dot's environment and reducing the impact of external noise sources, the emitted photons can have high temporal and spectral coherence, leading to high-fidelity quantum operations.

Frequently used Q-dots, which are self-assembled quantum dots, are created by well controlled and tedious techniques from semiconductor processing, including Epitaxial growing, layer-by-layer semiconductor self-assembly techniques. The process of self-assembled growth can form very small islands of smaller-band gap material surrounded by larger bandgap material and provides specific wavelength emission depending on the designed band gaps.

Quantum dot-based single-photon sources operate by exploiting the phenomenon of single-exciton recombination, where a single-electron hole pair (exciton) recombines and emits a single photon. The optical transitions in a Qdot can be excited by either electrical current injection or a continuous or pulsed laser. These transitions have nanosecond lifetimes. Therefore, the single-photon emission rates can be up to GHz and the highest compared to other systems. Complexity in the Qdot-based single-photon source arises from the requirement of cryogenic temperatures to reduce the inhomogeneous broadening in the optical transitions—this broadening results in reduced coherence and reduced purity of the single photons.

Typically, high-purity single-photon sources with Qdots have been realized by placing them in an optical cavity. This configuration enables on-demand single photons and their subsequent routing to the output channel via the cavity mode. Purcell-enhancement of Qdot emission into the cavity is also a crucial advantage of a cavity-based system. There are many ways in which a quantum dot is coupled to a cavity. Typical cavity structures are tunable microcavities, photonic crystal cavities, and micropillar cavities. The research works in [25, 20, 19] have focused on coupling a Qdot to a fiber-based cavity platform. Direct fiber output allows network integrable SPSs. In self-assembled Qdots, coupling the excitonic transitions to a fiber-based cavity has been shown in [20]. A DBR mirror under the Qdot layer is used as one of the mirrors of the cavity, Figure 4.9. By using a fiber-based mirror with an adjustable position perpendicular to the QD plane, one can have a Purcell-enhanced and guided single-photon generation into the cavity mode.

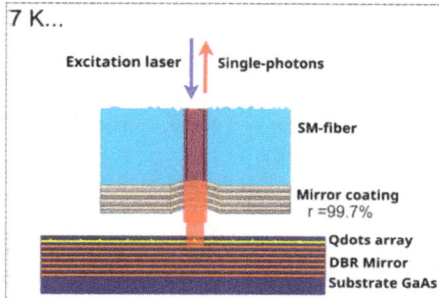

Figure 4.9: A fiber-based mirror is used to couple excitation light and guide the emitted single photon to and fro from a Qdot. The Qdots are grown over a GaAs substrate containing a distributed Bragg mirror underneath. The fiber mirror and the DBR mirror form a cavity to Purcell-enhanced single-photon emission [19, 20]. The Qdot is operated under cryogenic temperature, e. g., 7 Kelvin in [20]. This figure is adapted from [20], with the permission of AIP Publishing.

Several other techniques exist for coupling Qdots to photonic structures such as micropillar cavities and nanowires, enabling enhanced interaction and control over the emitted photons. These methods aim for the generation of high-quality single photons with improved indistinguishability. By continuous improvements on microresonators and coupling techniques of Qdots to the micro-resonators, researchers are actively working toward overcoming challenges associated with Qdot-based single-photon sources and toward the development of reliable and efficient systems for generating indistinguishable single photons.

4.4.2.2 NV centers coupled to a cavity

NV centers, or nitrogen vacancy centers, are atomic-scale defects in diamonds that possess remarkable optical and spin properties. They are highly researched systems for quantum technology because of very specific optical properties, such as

– single-photon emission at room temperatures,
– long-lived electronic and nuclear spin states allowing for quantum state manipulation and control for longer coherence time.

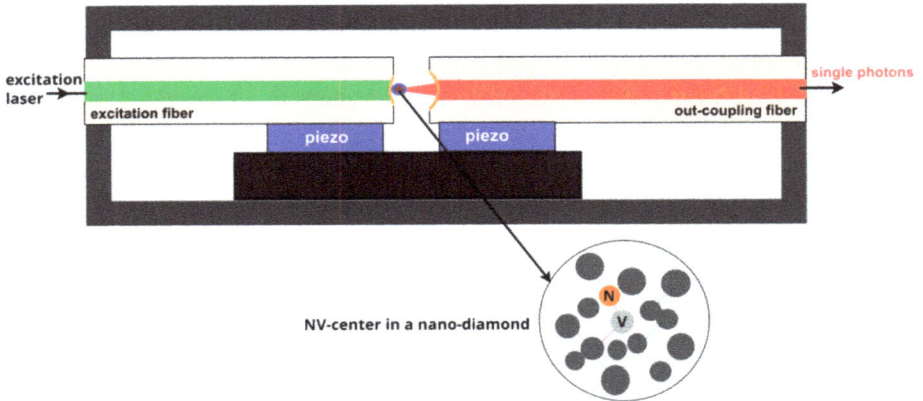

Figure 4.10: A schematic setup where a nanodiamond containing an NV center is directly embedded on one of the fiber mirror. One of the fibers is used for sending the laser to excite the transition at 532 nm wavelength. The other fiber collects a single photon emitted into the cavity. Piezos attached to the fiber mirrors allow the tuning of the cavity length close to the NV centers resonance.

NV color centers are formed inside a diamond lattice by a substitution of a carbon atom with a nitrogen atom, accompanied by a vacancy at an adjacent lattice position; see Figure 4.10. This defect structure gives rise to fascinating optical properties, containing a spin-triplet ground state and an excited state. The optical transitions between these states can be driven by light, allowing for the controlled generation of single photons. In a three-level energy system involving NV centers, there exists a ground state denoted as $|g\rangle$ and an excited state denoted as $|e\rangle$. The excited state $|e\rangle$ also has a thermal coupling to another metastable state denoted as g_2. However, since the state g_2 has a long lifetime, any population transfer through the excited state reduces the rate of single-photon emission. Therefore, it is desirable to minimize such thermal coupling and maximize the population transfer directly from the excited state to the ground state for efficient single-photon generation.

For the NV center, the emission line corresponds to a wavelength of 637 nm with a broad spectrum of 100 nm. One of the challenges in harnessing the emission from NV centers is the high refractive index of diamond ($n = 2.4$), which makes it difficult to extract/photons emitted from the bulk of the diamond. To overcome this limitation, diamond nanostructures are commonly employed. By incorporating NV centers into photonic crystals, nanowires, or disk resonators with fiber-based mirrors, one can take advantage of the Purcell effect, which enhances the emission rate into well-defined cavity modes. This allows for more efficient extraction and manipulation of the emitted photons from the NV centers. For example, by embedding NV centers in photonic crystals, nanowires, and disk resonators of fiber-based mirrors, one can channel the photon emission from the NV center directly into the fiber.

In references [1, 2], as shown in Figure 4.10, a nanodiamond is deposited on one of the fiber mirrors (or a plane mirror), and an another curved fiber mirror is used to guide the excitation beam at 532 nm. The photon emission from the nano-diamond is also collected via this fiber. Purcell enhancement can be observed in the photon emission rate into the fiber cavity.

The ability to operate NV center-based SPS at room temperature and their compatibility with diamond fabrication technologies make them promising candidates for practical quantum devices. Improvement and simplification of the techniques of depositing single nanodiamond on photonic structures will allow for high bandwidth single-photon sources at room temperature.

4.4.3 Single-photon characterization

Characterization of a single-photon source is crucial to assess the quality and performance of single photons, enabling optimization and comparison with desired specifications for specific applications in quantum communication, quantum information processing, and other quantum technologies. The characterization process involves several key measurements and techniques to assess its performance. Here are some common methods used to characterize a single-photon source:

- *Indistinguishability:* The Hong–Ou–Mandel (HOM) interference experiment is a powerful technique to characterize the indistinguishability of photons emitted from a single-photon source. It involves interfering with two photons on a balanced beam splitter at the input port and subsequently measuring the two output ports using photodetectors (Figure 4.11). The coincidence count rate of the detectors as a function of the arrival time delay indicates the photon's indistinguishability. A clear dip in the coincidence count rate at zero delays indicates the high indistinguishability of the photons.

- *Antibunching:* One of the fundamental characterizations of a single-photon source is determining its photon statistics. This involves measuring the probability distribution of detecting multiple photons at a given time interval. A high-quality single-photon source should exhibit sub-Poissonian statistics, indicating a low probability of emitting multiple photons simultaneously. Hanbury Brown–Twiss experiment allows to measure the photon antibunching behavior by measuring the $g^{(2)}(\tau)$ correlation function, which measures the probability of detecting two photons at different time intervals τ. For a perfect single-photon source, the $g^{(2)}(\tau)$ correlation function should show a dip at zero delays, indicating antibunching behavior, where photons are emitted one at a time. This measurement provides information about the indistinguishability of the emitted photons.

Figure 4.11 shows the experimental setup for both HBT and Hong–Ou–Mandel experiments. Typically single-photon counting detectors are used with associated electronics

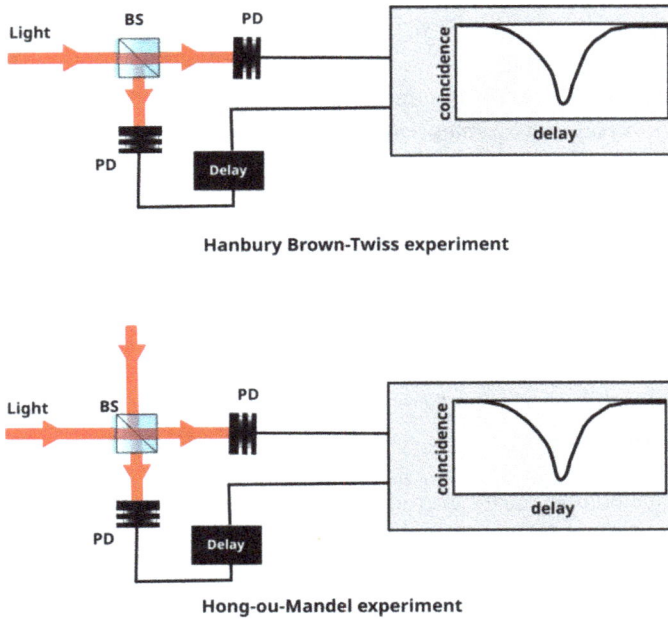

Figure 4.11: Experimental setup for photon characteristic measurements. BS denotes the beam splitter and PD is the photodetector.

for counting and evaluating the correlation between the registered events. As an example, an experimental result for photon antibunching effect is demonstrated in [9]. A single-neutral rubidium atom is strongly coupled to a FFPC. A excitation laser is used to drive a closed transition and emitted photons are collected using cavity mode. Purcell enhancement due to the cavity modifies the emission rate of the atom into the cavity. A HBT measurement shows a clear antibunching dip for zero delay between the detectors as shown in Figure 4.12.

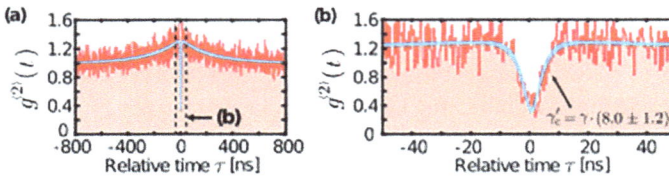

Figure 4.12: Experimental result demonstration the single-photon antibunching behavior [9]. A single-rubidium atom is used as a quantum emitter, which is strongly coupled to a FFPC. Figure (a) shows the antibunching dip at zero time delay between the detector. Figure (b) shows the zoomed view close to the antibunching dip. There is 8 times enhancement of the photon emission rate γ'_c into the cavity mode compared to the photon emission rate into free space γ. This figure is reprinted with permission from [9].

4.4.4 Quantum memories

Long-distance and distributed quantum information processing require strategies to overcome losses and any degradation of quantum states during the transfer of photonic quantum bits (qubits). These loss and decoherence compensation protocols involve converting flying qubits (photons) into long-lived stationary qubits (matter), and subsequently releasing them on demand. Quantum memories are devices or systems that are designed for this purpose to store and retrieve quantum states reliably and coherently. This can be achieved by mapping the quantum state of the flying qubit onto the internal states of the stationary qubit (matter) using techniques like quantum state transfer or entanglement swapping via controlled light-matter interaction [7]. The quantum memory is capable of holding the quantum state for extended periods.

When the quantum information is required at a remote location, the quantum memory can be read out or manipulated, on-demand, to retrieve the stored quantum state. This retrieval process allows for the conversion of the stationary qubit (stored quantum information) back into a flying qubit, which can be transmitted to the desired destination. The on-demand release ensures that the quantum state is only transferred when needed, reducing the chances of loss or degradation during transmission. Therefore, quantum memories (QMs) are one of the crucial components for future envisioned long-distance quantum networks and distributed quantum computing.

The desired features in quantum memories are long storage time, high efficiency, high fidelity, and scalability. An ideal quantum memory, in principle, has two conflicting requirements, which are relatively optimized for a realistic implementation and according to the requirement and constraints in a specific system:

- On one hand, quantum memories should allow perfect storage and retrieval of a photonic quantum state. This means it requires highly efficient interfacing of the photons to the matter;
- On the other hand, the information stored in a stationary qubit (stored quantum information) should be preserved against any decoherences for an arbitrarily long time. This means there should be no interaction of matter with the environment.

The storage time of a quantum memory is typically characterized by the coherence time or the dephasing time of the stored quantum state. Coherence time refers to the time scale over which the quantum state retains its phase information, while the dephasing time represents the time scale over which the coherence of the quantum state is lost. These times can be determined experimentally by measuring the decay of quantum coherence or by assessing the fidelity of state retrieval as a function of storage duration. A long coherence time is desired to synchronize the probabilistic events in the two or more different network links. Synchronization means storing qubits and retiming them until all links are ready. This is important in quantum repeaters and quantum computing with linear optics [5, 22, 15, 14]. There have been various implementations of quantum

Table 4.1: Few representative examples of quantum memory systems.The references included in the table employ diverse approaches to quantify memory efficiencies. Storage efficiency refers solely to the efficiency of photon storage within the atom, while retrieval efficiency encompasses the total efficiency achieved over the memory's lifespan. End-to-end efficiency additionally accounts for all losses in the system.

Quantum Memories					
QM-system	Storage time	Efficiency	Fidelity	Banwidth	Reference
Single-atom in Fiber-cavity					
^{87}Rb atom	170 µs	56 % retrieval	94.7 %	60 MHz	[4]
^{87}Rb atom	1 µs	8 % storage	–	5 ns pulse	[17]
^{171}Yb^{+} ion	1 ms	10.1 % retrieval	–	19 MHz	[30]
Color-centers in Diamond					
Nitrogen-vacancy	10 s	1 %	98 %	12 MHz	[31]
Silicon-vacancy	2 s	14 %	95 %	16.7 ns pulse	[27]
Cold-atomic ensemble					
Cesium	15 µs	85 %	–	4.5 MHz	[6]
Rubidium	220 ms	76 %	–	–	[32]
Warm atomic vapor					
Rubidium	50 ns	3.4 % end-to-end	–	660 MHz	[29]
Rubidium	100 ns	25 % end-to-end	–	1.7 ns pulse	[8]
Rubidium cavity-enhanced	1.2 µs	67 %	97 %	–	[16]
Cesium	1-2 µs	–	–	500 MHz	[21]
Cesium atoms	5.4 ns	5 % end-to-end	–	1 GHz	[13]

memories with the pros and cons of each system. Table 4.1 presents some of the current QM implementations.

The fidelity of a quantum memory refers to the measure of how well a stored quantum state can be retrieved or reproduced when it is read out or accessed from the memory. It quantifies the similarity or closeness between the original quantum state that was stored and the retrieved state. The efficiency of a quantum memory refers to the measure of how effectively quantum information can be stored and retrieved in the memory system. Both fidelity and efficiency of a quantum memory help in quantifying the fraction of input quantum information that is successfully stored and later retrieved without significant loss or degradation.

Another important feature required for QMs is scalability and also fiber-based interfaces. Direct fiber-coupled quantum memories are essential for their integration in fiber-based networks and are relevant in the context of this book.

4.4.4.1 Optical quantum memories

One of the typical realizations of optical quantum memories is based on λ-type energy configuration in a matter-based system. In this configuration, an excited state $|e\rangle$ is cou-

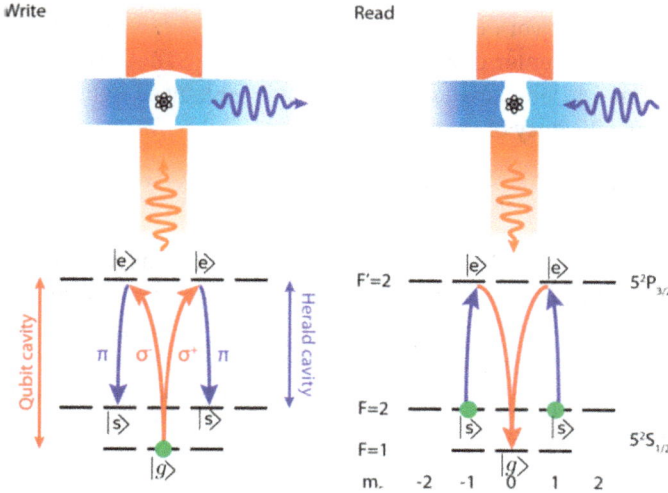

Figure 4.13: Experimental setup in [4] used two crossed fiber cavities simultaneously coupled to a single-rubidium atom in strong-coupling regime. Shown in red, the Qubit cavity, supports c^+/σ^- transition, while blue, Herlad cavity, support π-transition. During memory storage process, circularly polarized photon (σ^+/σ^-) is absorbed into the atom via Qubit cavity. At the same time, a single photon is emitted along the herald cavity due to the vacuum STIRAP process. Memory read out is the time reversal process of the memory storage. The lower figure demonstrates the relevant energy level of the atom. For the configuration used in the experiment, two polarization degenerate λ-system ((σ^+,π), (σ^-,π)) are used for the realization of quantum memory for polarized photonic quantum bits. This figure is reprinted from [4] with permission from SNCSC.

pled to a ground state $|g\rangle$ and a long-lived metastable state $|s\rangle$ via dipole transitions. Typically, forbidden $|g\rangle \rightarrow |s\rangle$ allows long coherence time while in the storage state $|s\rangle$.

As shown in lower Figure 4.13, coherent absorption and retrieval of the photonic state are mediated by the STIRAP process, as discussed in the previous Section 4.3.1. In the photon storage scheme, a photon pulse arrives such that it addresses the dipole transition $|g\rangle \rightarrow |e\rangle$, while a strong control pulse is applied to address $|s\rangle \rightarrow |e\rangle$ transition. Both light fields are in two photon-resonance conditions for $|g\rangle \rightarrow |s\rangle$:

$$\omega_{ge} - \omega_{es} = \omega_{gs} \qquad (4.43)$$

By controlling the detuning of the two fields and the temporal shaping of the control pulse, photon storage and retrieval efficiencies can be maximized.

In a cavity-enhanced quantum memory scheme, photon storage is enhanced by using a high finesse cavity along the $|g\rangle \rightarrow |e\rangle$ transition of the λ-configuration. The control pulse can also be enhanced by using another cavity. However, the high finesse of the cavity along the weak photon signal side enhances the coherent photon cavity absorption.

Fiber-based resonators interacting with single atoms/ions, nitrogen vacancy centers, quantum dots, or rare-earth ions are the currently researched systems with a promise toward realizing fiber network compatible quantum memories. For example, for a single atom trapped at the center of a fiber-cavity mode, the STIRAP process is used for storing the photon and on-demand retrieval. Cavity-enhanced light-matter interaction enables the high efficiency of the process given by [10],

$$\eta = \frac{C}{1+C}. \tag{4.44}$$

One can obtain higher efficiencies for a given quantum system by having high finesse cavities with small mode volumes. However, smaller mode volume cavities are shorter in length and increasingly obstruct the optical access of the matter inside. Fiber cavities, in this regard, have shown great traction due to their:
-   small mode volume, typical cavity length ranging from few 10–100 μm and mode waist from few μm to few tens of μm;
-   open and easy access;
-   direct fiber-coupled input and output channels.

Reference [17] has used a fiber-based cavity in the strong coupling regime to store photon pulses into a single atom trapped at the center of the cavity. Reference [4] has used two crossed fiber-based cavities to store and retrieve polarization-based qubits in Rubidium atoms (see Figure 4.13). A typical implementation of quantum memory in a single-atom cavity system involves the interaction between a single atom and a cavity mode. This setup allows for the storage and retrieval of quantum information in the atom's internal states (λ-configuration) while utilizing the cavity mode as a resource for coherent storage. The following is a step-by-step description of the implementation:
-   The system starts with a laser cooling and trapping of a single atom. The atomic system often has well-defined internal energy levels, such as λ-configuration with two ground states and an excited state. Common examples include atoms like cesium (Cs), rubidium (Rb), or trapped ions.
-   The atom is placed inside an optical cavity, which consists of two highly reflective mirrors that form a resonant cavity. In the case of a nanofiber cavity or WGM resonator, the atom is coupled via the evanescent field when placed closed to the resonator. The cavity enhances the interaction between the atom and the electromagnetic field, enabling the efficient transfer of quantum information.
-   Initialization: The atom is prepared in a specific internal state, usually the ground state, ready to receive quantum information and coupled to the cavity.
-   Storage: The quantum information to be stored is typically in the form of a coherent optical pulse or a single photon. This information is encoded into the atomic system by coupling the optical pulse to the atom-cavity system. Using the STIRAP method, a control pulse adiabatically controls the photonic qubit pulse into the atom-cavity system. Therefore, the quantum information is transferred from the optical mode to the internal energy levels of the atom. The cavity mode acts as a "buffer" to hold the quantum information coherently in the atom.

During the storage period, the quantum information is preserved in the atom's internal states, which can be long-lived compared to the coherence time of the cavity mode. The atom-cavity system is carefully controlled to maintain the coherence of the stored quantum state.

－ Retrieval: When the stored quantum information is desired, a retrieval process is initiated. This process is typically the reverse of the storage process. The atom-cavity system is manipulated to transfer the quantum information from the atom's internal states back into the optical mode by applying a time reversal form of the control pulse used during the storage process. The retrieved quantum information is encoded onto an optical pulse or photon, which can be transmitted to other quantum devices or used for subsequent quantum processing tasks.

The fidelity and efficiency of the quantum memory are assessed by comparing the retrieved state with the original input state. Techniques such as quantum state tomography or interference measurements can be used to verify the quality of the stored and retrieved states.

Bibliography

[1] R. Albrecht, A. Bommer, C. Deutsch, J. Reichel, and C. Becher. Coupling of a single nitrogen-vacancy center in diamond to a fiber-based microcavity. Physical Review Letters, 110:243602, Jun 2013. https://doi.org/10.1103/PhysRevLett.110.243602. URL https://link.aps.org/doi/10.1103/PhysRevLett.110.243602.

[2] R. Albrecht, A. Bommer, C. Pauly, F. Mücklich, A. W. Schell, P. Engel, T. Schröder, O. Benson, J. Reichel, and C. Becher. Narrow-band single photon emission at room temperature based on a single nitrogen-vacancy center coupled to an all-fiber-cavity. Applied Physics Letters, 105(7):073113, 08 2014. ISSN 0003-6951. https://doi.org/10.1063/1.4893612.

[3] Y. Arakawa and M. J. Holmes. Progress in quantum-dot single photon sources for quantum information technologies: A broad spectrum overview. Applied Physics Reviews, 7(2):021309, 06 2020. ISSN 1931-9401. https://doi.org/10.1063/5.0010193.

[4] M. Brekenfeld, D. Niemietz, J. D. Christesen, and G. Rempe. A quantum network node with crossed optical fibre cavities. Nature Physics, 16(6):647–651, Jun 2020. ISSN 1745-2481. https://doi.org/10.1038/s41567-020-0855-3.

[5] H-J. Briegel, W. Dür, J. I. Cirac, and P. Zoller. Quantum repeaters: the role of imperfect local operations in quantum communication. Physical Review Letters, 81(26):5932, 1998.

[6] M. Cao, F. Hoffet, S. Qiu, A. S. Sheremet, and J. Laurat. Efficient reversible entanglement transfer between light and quantum memories. Optica, 7(10):1440–1444, Oct 2020. https://doi.org/10.1364/OPTICA.400695. URL https://opg.optica.org/optica/abstract.cfm?URI=optica-7-10-1440.

[7] J. I. Cirac, P. Zoller, H. J. Kimble, and H. Mabuchi. Quantum state transfer and entanglement distribution among distant nodes in a quantum network. Physical Review Letters, 78:3221–3224, Apr 1997. https://doi.org/10.1103/PhysRevLett.78.3221. URL https://link.aps.org/doi/10.1103/PhysRevLett.78.3221.

[8] O. Davidson, O. Yogev, E. Poem, and O. Firstenberg. Single-photon synchronization with a room-temperature atomic quantum memory. Physical Review Letters, 131:033601, Jul 2023. https://doi.org/10.1103/PhysRevLett.131.033601. URL https://link.aps.org/doi/10.1103/PhysRevLett.131.033601.

[9] J. Gallego, W. Alt, T. Macha, M. Martinez-Dorantes, D. Pandey, and D. Meschede. Strong purcell effect on a neutral atom trapped in an open fiber cavity. *Physical Review Letters*, 121:173603, Oct 2018. https://doi.org/10.1103/PhysRevLett.121.173603. URL https://link.aps.org/doi/10.1103/PhysRevLett.121.173603.

[10] A. V. Gorshkov, A. André, M. D. Lukin, and A. S. Sørensen. Photon storage in Λ-type optically dense atomic media. i. cavity model. *Physical Review A*, 76:033804, Sep 2007. https://doi.org/10.1103/PhysRevA.76.033804. URL https://link.aps.org/doi/10.1103/PhysRevA.76.033804.

[11] S. Haroche. Nobel lecture: Controlling photons in a box and exploring the quantum to classical boundary. *Reviews of Modern Physics*, 85:1083–1102, Jul 2013. https://doi.org/10.1103/RevModPhys.85.1083. URL https://link.aps.org/doi/10.1103/RevModPhys.85.1083.

[12] M. Hijlkema, B. Weber, H. P. Specht, S. C. Webster, A. Kuhn, and G. Rempe. A single-photon server with just one atom. *Nature Physics*, 3(4):253–255, Apr 2007. ISSN 1745-2481. https://doi.org/10.1038/nphys569.

[13] K. T. Kaczmarek, P. M. Ledingham, B. Brecht, S. E. Thomas, G. S. Thekkadath, O. Lazo-Arjona, J. H. D. Munns, E. Poem, A. Feizpour, D. J. Saunders, J. Nunn, and I. A. Walmsley. High-speed noise-free optical quantum memory. *Physical Review A*, 97:042316, Apr 2018. https://doi.org/10.1103/PhysRevA.97.042316. URL https://link.aps.org/doi/10.1103/PhysRevA.97.042316.

[14] P. Kok, W. J. Munro, K. Nemoto, T. C. Ralph, J. P. Dowling, and G. J. Milburn. Linear optical quantum computing with photonic qubits. *Reviews of Modern Physics*, 79:135–174, Jan 2007. https://doi.org/10.1103/RevModPhys.79.135. URL https://link.aps.org/doi/10.1103/RevModPhys.79.135.

[15] A. I. Lvovsky, B. C. Sanders, and W. Tittel. Optical quantum memory. *Nature Photonics*, 3(12):706–714, Dec 2009. ISSN 1749-4893. https://doi.org/10.1038/nphoton.2009.231.

[16] L. Ma, X. Lei, J. Yan, R. Li, T. Chai, Z. Yan, X. Jia, C. Xie, and K. Peng. High-performance cavity-enhanced quantum memory with warm atomic cell. *Nature Communications*, 13(1):2368, May 2022. ISSN 2041-1723. https://doi.org/10.1038/s41467-022-30077-1.

[17] T. Macha, E. Uruñuela, W. Alt, M. Ammenwerth, D. Pandey, H. Pfeifer, and D. Meschede. Nonadiabatic storage of short light pulses in an atom-cavity system. *Physical Review A*, 101:053406, May 2020. https://doi.org/10.1103/PhysRevA.101.053406. URL https://link.aps.org/doi/10.1103/PhysRevA.101.053406.

[18] J. McKeever, A. Boca, A. D. Boozer, R. Miller, J. R. Buck, A. Kuzmich, and H. J. Kimble. Deterministic generation of single photons from one atom trapped in a cavity. *Science*, 303(5666):1992–1994, 2004. https://doi.org/10.1126/science.1095232. URL https://www.science.org/doi/abs/10.1126/science.1095232.

[19] J. Miguel-Sánchez, A. Reinhard, E. Togan, T. Volz, A. Imamoglu, B. Besga, J. Reichel, and J. Estève. Cavity quantum electrodynamics with charge-controlled quantum dots coupled to a fiber Fabry–Perot cavity. *New Journal of Physics*, 15(4):045002, Apr 2013. https://doi.org/10.1088/1367-2630/15/4/045002.

[20] A. Muller, E. B. Flagg, M. Metcalfe, J. Lawall, and G. S. Solomon. Coupling an epitaxial quantum dot to a fiber-based external-mirror microcavity. *Applied Physics Letters*, 95(17):173101, 10 2009. ISSN 0003-6951. https://doi.org/10.1063/1.3245311.

[21] X. -L. Pang, A. -L. Yang, J. -P. Dou, H. Li, C. -N. Zhang, E. Poem, D. J. Saunders, H. Tang, J. Nunn, I. A. Walmsley, and X. -M. Jin. A hybrid quantum memory–enabled network at room temperature. *Science Advances*, 6(6):eaax1425, 2020. https://doi.org/10.1126/sciadv.aax1425. URL https://www.science.org/doi/abs/10.1126/sciadv.aax1425.

[22] S. Pirandola, J. Eisert, C. Weedbrook, A. Furusawa, and S. L. Braunstein. Advances in quantum teleportation. *Nature Photonics*, 9(10):641–652, 2015.

[23] E. M. Purcell. Proceedings of the american physical society. *Physical Review*, 69:681, Jun 1946. https://doi.org/10.1103/PhysRev.69.674. URL https://link.aps.org/doi/10.1103/PhysRev.69.674.

[24] A. Reiserer and G. Rempe. Cavity-based quantum networks with single atoms and optical photons. *Reviews of Modern Physics*, 87:1379–1418, Dec 2015. https://doi.org/10.1103/RevModPhys.87.1379. URL https://link.aps.org/doi/10.1103/RevModPhys.87.1379.

[25] H. Snijders, J. A. Frey, J. Norman, V. P. Post, A. C. Gossard, J. E. Bowers, M. P. van Exter, W. Löffler, and D. Bouwmeester. Fiber-coupled cavity-qed source of identical single photons. *Physical Review Applied*, 9:031002, Mar 2018. https://doi.org/10.1103/PhysRevApplied.9.031002. URL https://link.aps.org/doi/10.1103/PhysRevApplied.9.031002.

[26] N. Somaschi, V. Giesz, L. De Santis, J. C. Loredo, M. P. Almeida, G. Hornecker, S. L. Portalupi, T. Grange, C. Antón, J. Demory, C. Gómez, I. Sagnes, N. D. Lanzillotti-Kimura, A. Lemaítre, A. Auffeves, A. G. White, L. Lanco, and P. Senellart. Near-optimal single-photon sources in the solid state. *Nature Photonics*, 10(5):340–345, May 2016. ISSN 1749-4893. https://doi.org/10.1038/nphoton.2016.23.

[27] P. -J. Stas, Y. Q. Huan, B. Machielse, E. N. Knall, A. Suleymanzade, B. Pingault, M. Sutula, S. W. Ding, C. M. Knaut, D. R. Assumpcao, Y. -C. Wei, M. K. Bhaskar, R. Riedinger, D. D. Sukachev, H. Park, M. Lončar, D. S. Levonian, and M. D. Lukin. Robust multi-qubit quantum network node with integrated error detection. *Science*, 378(6619):557–560, 2022. https://doi.org/10.1126/science.add9771. URL https://www.science.org/doi/abs/10.1126/science.add9771.

[28] N. V. Vitanov, A. A. Rangelov, B. W. Shore, and K. Bergmann. Stimulated raman adiabatic passage in physics, chemistry, and beyond. *Reviews of Modern Physics*, 89:015006, Mar 2017. https://doi.org/10.1103/RevModPhys.89.015006. URL https://link.aps.org/doi/10.1103/RevModPhys.89.015006.

[29] J. Wolters, G. Buser, A. Horsley, L. Béguin, A. Jöckel, J. Jahn, R. Warburton, and P. Treutlein. Simple Atomic Quantum Memory Suitable for Semiconductor Quantum Dot Single Photons. *Phys. Rev. Lett.*, 119:060502, Aug 2017. https://doi.org/10.1103/PhysRevLett.119.060502. URL https://link.aps.org/doi/10.1103/PhysRevLett.119.060502.

[30] P. Kobel, M. Breyer, and M. Köhl. Deterministic spin-photon entanglement from a trapped ion in a fiber Fabry–Perot cavity. *npj Quantum Information*, 7:1, Jan 2021. https://doi.org/10.1038/s41534-020-00338-2.

[31] S. Yang, Y. Wang, D. D. B. Rao, T. H. Tran, A. S. Momenzadeh, M. Markham, D. J. Twitchen, P. Wang, W. Yang, R. Stöhr, P. Neumann, H. Kosaka, and J. Wrachtrup. High-fidelity transfer and storage of photon states in a single nuclear spin. *Nature Photonics*, 10:8, Aug 2016. https://doi.org/10.1038/nphoton.2016.103.

[32] S. -J. Yang, X. -J. Wang, X. -H. Bao, and J. -W. Pan. An efficient quantum light–matter interface with sub-second lifetime. *Nature Photonics*, 10:6, Jun 2016. https://doi.org/10.1038/nphoton.2016.51.

5 Current and future quantum technology with fiber resonators

Quantum technology is rapidly evolving as a groundbreaking field with the potential to revolutionize various areas of science, computing, communication, and sensing. At the heart of many quantum systems and devices are resonator-based light-matter interfaces, which are crucial for manipulating and controlling quantum states.

Photons (light), due to their exceptional ability to transport quantum information, serve as excellent carriers, while matter is essential for storing and processing photonic quantum bits. Consequently, the development of efficient light-matter interfaces emerges as a critical component within the realm of quantum technology. In this context, fiber-based resonators have been researched in diverse areas of quantum technology. One prominent application is in cavity quantum electrodynamics (CQED), where the interaction between light and matter in confined spaces is explored to achieve enhanced control and manipulation of quantum states. Fiber-resonators provide a platform for CQED experiments, facilitating strong light-matter interactions and enabling the development of quantum memory devices. These devices may become crucial for quantum networks and distributed quantum information processing.

In this concluding chapter, we will provide a brief overview of proven instances where fiber-based resonators have been employed to advance quantum technological applications and explore their potential roles in shaping future quantum networks. We will first elucidate the concept of distributed quantum information processing, followed by examining quantum networks centered around fiber resonators.

5.1 Distributed quantum information processing

In a traditional quantum information processing scheme, a single centralized quantum computer performs computations on a quantum processor. Presently, the computing potential of quantum processors is significantly constrained due to limitations associated with scaling the number of quantum gate operations. Although ongoing research and anticipated advancements hold promise for enhancing quantum processing capacity and system scaling, there are always some constraints on the largest attainable system size of a single processor during any development stage, limiting the quantum computation capacity. Distributed quantum information processing promises to enhance the quantum information processing capabilities by distributing the computational workload across multiple quantum nodes that are connected through quantum networks, (see Figure 5.1). This distributed approach offers several advantages, including increased scalability, fault tolerance, and the ability to leverage resources from geographically dispersed quantum systems. One of the primary motivations for distributed quantum information processing is to overcome the limitations of local quantum systems, such

https://doi.org/10.1515/9783110636260-005

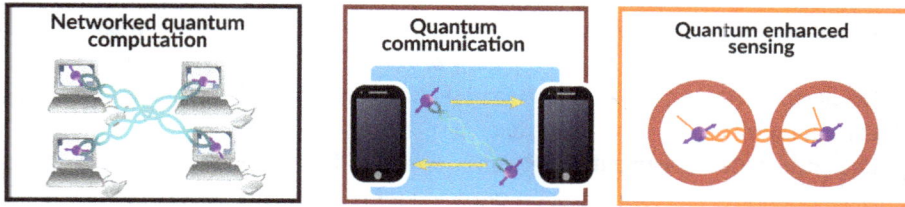

Figure 5.1: Illustrative cartoons depicting distributed-quantum networks, envisioned for various tasks under quantum advantage. The quantum states, symbolized by spherical arrows, are entangled over vast distances and distributed across different locations to realize a distributed quantum network. These networks can be used for quantum computation, communication, or sensing.

as limited qubit resources and susceptibility to errors. By distributing the processing tasks, it becomes possible to harness the combined power of multiple quantum nodes, effectively increasing the computational capacity, enabling the execution of large and complex quantum algorithms, and ultimately realizing quantum internet [28, 16].

5.1.1 Distributed networks

Distributed communication networks are networks in which multiple nodes or devices are interconnected to enable the exchange of information and communication over a distributed or decentralized architecture. Unlike centralized networks, where communication flows through a central entity, distributed networks distribute the communication load across multiple nodes, enhancing scalability, fault tolerance, and efficiency.

In distributed communication networks, each node has its own processing and communication capabilities, allowing them to communicate directly with neighboring nodes without relying on a central authority. These networks can be structured in various topologies, such as mesh, ring, tree, or random connections, depending on the specific requirements and intent of the application.

Distributed communication networks offer several advantages, including:

- Scalability: Distributed networks can easily scale by adding or removing nodes without significantly impacting the overall network performance. This scalability is particularly useful in large-scale systems or in situations where the network size may change dynamically.
- Fault tolerance: Distributed networks are inherently more robust against failures or disruptions. If a node fails or a link is broken, the network can reroute communication through alternative paths, ensuring the continuity of communication. This fault tolerance makes distributed networks more resilient to single-point failures and improves overall system reliability.
- Performance and efficiency: By distributing the communication load among multiple nodes, distributed networks can improve overall network performance and reduce congestion. Communication can occur simultaneously between multiple pairs of nodes, enabling efficient data transfer and reducing latency.

- Flexibility and adaptability: Distributed networks can be easily reconfigured and adapted to changes in network topology or requirements. Nodes can join or leave the network in a dynamic fashion, and the network can self-organize and reestablish communication paths as needed.
- Decentralization: Distributed networks promote decentralization, as there is no central authority or control point governing the network. This decentralized nature allows for greater autonomy, peer-to-peer communication, and collaborative decision-making among nodes.

Distributed communication networks [20] find applications in various domains, including telecommunications, Internet of Things (IoT), peer-to-peer networks, wireless sensor networks, and distributed computing. Examples of distributed communication networks include ad hoc networks, wireless mesh networks, blockchain networks, and decentralized internet protocols like InterPlanetary File Systems (IPFS).

However, distributed communication networks also pose challenges, such as ensuring network security, managing network resources effectively, maintaining synchronization and coordination among nodes, and optimizing routing and data transfer algorithms. Nonetheless, the benefits of distributed communication networks make them well suited for scenarios that require scalability, fault tolerance, and efficient communication in a decentralized environment.

A quantum version of the distributed network relies on a specialized type of distributed communication network that leverages the principles and properties of quantum mechanics to enable secure and efficient communication and information processing. These networks harness the unique features of quantum systems, such as superposition, entanglement, and quantum interference, to perform tasks that are not feasible with classical communication networks.

5.1.2 Distributed quantum networks

One of the key advantages of the quantum version of the distributed networks is the ability to achieve secure communication through quantum cryptography protocols. As an illustration, in the most basic setup of a two-node quantum network, where quantum key distribution (QKD) is employed, it becomes possible to establish a shared secret key between two parties through the utilization of quantum entanglement [11]. This key can then be used for secure communication, as any eavesdropping attempt would disturb the entanglement and be detectable. QKD offers a provably secure method of key distribution, ensuring the confidentiality of transmitted data [2, 30].

Quantum distributed networks can also enable distributed quantum computation, where multiple quantum computers collaborate to solve complex computational problems. This is achieved through quantum entanglement and quantum teleportation, which allow for the transfer and manipulation of quantum states between distributed nodes. By distributing computational tasks across multiple quantum devices, distributed quantum networks can potentially enhance computational power and enable the solution of problems that are beyond the capabilities of individual quantum computers.

Moreover, quantum distributed networks can facilitate distributed quantum sensing and metrology, where quantum systems are used to enhance the precision and sensitivity of measurements. Quantum sensors distributed across a network can collectively gather information and perform measurements with higher accuracy than classical sensors.

However, building practical quantum-distributed networks face significant challenges. Quantum systems are extremely delicate and susceptible to noise and decoherence, which can degrade the quality of transmitted qubits. Developing reliable quantum channels, establishing and maintaining entanglement between nodes, and mitigating errors and noise are ongoing research areas in quantum networking [28].

Despite the challenges, quantum distributed networks hold great promise for revolutionizing communication, computation, and sensing technologies. They have the potential to enable secure communication, enhance computational capabilities, and advance scientific and technological applications that require quantum information processing over distributed architectures.

5.1.3 Quantum repeater protocols

The fundamental challenge in long-distance quantum communication is the loss of quantum information during transmission through optical fibers or other mediums. As the distance increases, the probability of successfully transmitting quantum states decreases due to photon absorption and scattering. Quantum repeater protocols address this issue by breaking the long-distance transmission into smaller segments and implementing entanglement swapping and purification techniques at each repeater node. Therefore, they aim to overcome the limitations of distance and loss in transmitting quantum information over long distances by establishing entanglement between distant nodes through a series of intermediate nodes called repeaters. These repeaters effectively extend the range of entanglement and enable reliable and efficient quantum communication by controlled light-matter interactions.

The basic concept of quantum repeater protocols involves the following steps:
- Entanglement generation: At the initial nodes, pairs of entangled qubits (often photons) are created using methods like spontaneous parametric down-conversion or single-photon sources. These entangled pairs are distributed to intermediate repeater nodes.
- Entanglement swapping: At each intermediate repeater node, entanglement swapping is performed. This involves combining entangled pairs from adjacent segments and using a Bell measurement to create entanglement between the pairs from non-adjacent segments. Bell measurements involve combining single photons from two nodes at a beam splitter's input ports and using two detectors at the output ports to

eliminate which-way information, thereby generating entanglement. This process enables the extension of entanglement across longer distances.

–   Entanglement purification: After entanglement swapping, the entangled pairs may be subject to noise and errors. Entanglement purification protocols are employed to enhance the quality of entanglement by removing errors and imperfections. These protocols involve comparing measurement outcomes between entangled pairs and applying appropriate operations to improve the fidelity of entanglement.

–   Repeater node operations: At each repeater node, quantum operations are performed on the qubits to enable entanglement swapping and purification. This typically involves photon measurements, entangling gate operations, and classical communication between nodes to coordinate the process.

By repeating the entanglement generation, entanglement swapping, and entanglement purification steps iteratively along the communication path, quantum repeater protocols enable the establishment of high-quality entanglement between distant nodes, overcoming the limitations of loss and distance in quantum communication.

Several specific protocols have been proposed for quantum repeaters, including the original quantum repeater protocol by Briegel et al. (1998) and the more recent DLCZ (Duan–Lukin–Cirac–Zoller) protocol. These protocols vary in their specific implementation details, but they share the common goal of extending entanglement over long distances by employing entanglement swapping and purification techniques.

Quantum repeater protocols are an active area of research and ongoing efforts are focused on improving the efficiency, fidelity, and scalability of these protocols to enable practical long-distance quantum communication and networking applications.

5.1.3.1 BDCZ protocol

The quantum repeater protocol proposed by Briegel, Dür, Cirac, and Zoller in 1998 is a pioneering approach to enable long-distance quantum communication known as the "BDCZ protocol" (referring to the initials of the authors) [7]. This protocol addresses the challenge of transmitting quantum information over long distances by mitigating the detrimental effects of decoherence and loss that occur during photon transmission.

The BDCZ protocol combines the concept of entanglement swapping and quantum memories to extend the achievable distances for quantum communication. The entanglement over long distances is established between distant qubits by connecting smaller segments with intermediate entangled qubits using the entanglement swapping technique. The protocol consists of the following key steps:

–   Entanglement generation: Initially, small clusters of qubits, referred to as "quantum repeater nodes," are distributed along the communication channel. Each node consists of a few qubits that can be atoms, ions, or solid-state systems. Within each node, local entanglement is generated between adjacent qubits using techniques such as laser cooling and trapping, or electron spin manipulation.

- Entanglement swapping: Once entanglement is created within neighboring nodes, entanglement swapping is performed between adjacent nodes to extend the entanglement over longer distances. This is achieved by measuring a pair of qubits in neighboring nodes in a specific entangled basis. The measurement outcome is then communicated to the other pair of qubits in the respective nodes, which induces entanglement between them. This swapping process can be repeated iteratively, allowing entanglement to be extended over multiple nodes.
- Quantum communication: After entanglement swapping, long-distance entanglement is established between distant nodes in the network. This entanglement can be used for quantum communication tasks such as quantum teleportation or secure quantum key distribution. To transmit quantum information, the sender performs a Bell measurement on their local qubit and the entangled qubit at the sender's node. The measurement outcome is then communicated classically to the receiver, who applies the appropriate quantum operation on their local qubit based on the received information, effectively teleporting the quantum state.

By utilizing entanglement swapping between adjacent nodes, the BDCZ protocol enables the distribution of entanglement over longer distances than direct transmission of photons. The iterative nature of the protocol allows for the creation of "entanglement links" between distant nodes, significantly increasing the communication range.

The BDCZ protocol laid the foundation for subsequent developments in quantum repeater technology. While the original protocol focused on entanglement swapping, subsequent research has explored additional techniques to enhance the protocol's efficiency, such as entanglement purification and quantum error correction codes. These advancements aim to improve the quality and fidelity of the transmitted entanglement, enabling long-distance quantum communication with high reliability. It is important to note that the BDCZ protocol, like other quantum repeater protocols, still faces challenges in terms of physical implementation, scalability, and resource requirements. However, it remains a significant contribution to the field of quantum communication and serves as a basis for ongoing research and development in the quest for practical quantum repeater systems.

5.1.3.2 DLCZ protocol

The DLCZ (Duan–Lukin–Cirac–Zoller) protocol is a quantum repeater protocol proposed by Duan, Lukin, Cirac, and Zoller in 2001 [10]. It provides a method for long-distance entanglement distribution, allowing the creation of entanglement between distant nodes in a quantum communication network. The DLCZ protocol leverages the interaction between photons and atomic ensembles to create and distribute entanglement. The use of atomic ensembles as quantum memories allows for efficient storage and retrieval of quantum information. By employing entanglement swapping and connection techniques, the protocol enables the distribution of entanglement over longer distances than

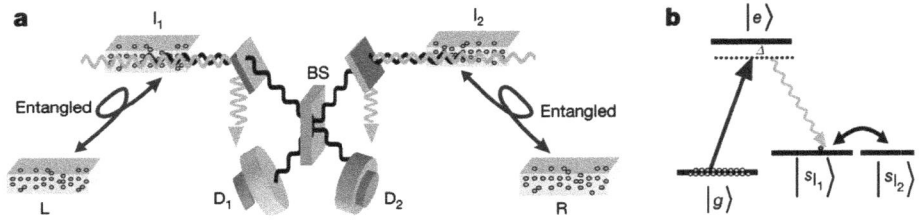

Figure 5.2: A schematic depiction of the conceptual setup for entanglement distribution and entanglement swapping. As an illustrative example, atomic ensembles and photons are used to create the collective entangled state here. The atomic ensembles have λ-type energy level configuration, as shown in Figure (b). In the first step, the ensemble pairs (L, l_1) and (R, l_2) are entangled by sharing a photon excitation between them. In the second step, the atomic excitation from ensembles l_1 and l_2 is converted simultaneously into light by applying an excitation pulse between respective transitions $|s\rangle \longrightarrow |e\rangle$. The emitted photon is then mixed on a symmetric beam splitter and detected on a two-detector configuration. In an ideal scenario, if either of the detectors clicks for single-photon detection, then the atomic ensemble on the extreme ends, i. e., L and R get entangled. Therefore, entanglement swapping takes place from entangled pairs (L, l_1) and (R, l_2) to L, R. This figure is reprinted from [10] with permission © Springer Nature.

direct photon transmission (Figure 5.2). The DLCZ protocol is based on the use of atomic ensembles as quantum memory and the controlled emission and absorption of photons by these ensembles. The basic idea of the protocol involves the following steps:

– Entanglement generation: Initially, two atomic ensembles are prepared, each consisting of a collection of atoms with suitable energy levels. These ensembles are placed at distant locations. Within each ensemble, the atoms are collectively coupled, for example, to a single cavity mode. By applying suitable control fields, the ensembles are prepared in a specific state called a "spin wave" state, which represents collective atomic excitations.

– Entanglement swapping: At each location, a single photon is prepared in a state called a "flying qubit." These flying qubits are sent from one ensemble to the other. Upon arrival, the flying qubits interacts with the local ensemble through a controlled absorption process. This interaction transfers the quantum state of the flying qubits onto the collective excitation of the local ensemble.

– Entanglement connection: After the absorption process, the ensembles are entangled with each other. This entanglement is achieved by creating a correlation between the collective excitations of the two ensembles. The entanglement connection is established through a process called quantum state transfer, which involves controlled emission and absorption of photons by the ensembles.

– Entanglement distribution: The entanglement connection between the ensembles allows for the distribution of entanglement to other nodes in the network. Additional flying qubits can be prepared and sent to the ensembles, and the entanglement swapping and connection steps can be repeated to generate entanglement between the ensembles and the new flying qubits. This enables the extension of entanglement over long distances.

The DLCZ protocol has been a significant advancement in the field of quantum repeaters, as it provides a feasible method for long-distance entanglement distribution. It has inspired further research and development in the field, leading to improvements and variations of the protocol. However, it is important to note that the DLCZ protocol, like other quantum repeater protocols, faces practical challenges such as the need for precise control of quantum states, efficient photon-ensemble interactions, and minimizing decoherence and noise. Ongoing research aims to address these challenges and enhance the performance and scalability of the DLCZ protocol for real-world quantum communication applications. Quantum memories, the essential components for quantum repeater protocols, are also realized by fiber-resonator-based CQED systems, as discussed in Chapter 4.

5.2 Experimental approaches for quantum networks

Experimental efforts in distributed quantum networks focus on building reliable quantum links over long distances, implementing efficient quantum repeaters, developing quantum memories with extended coherence times and demonstrating secure quantum communication protocols. Researchers are actively exploring various system combinations, advanced quantum error correction codes, and sophisticated quantum control techniques to overcome technical challenges and realize practical distributed quantum networks capable of performing complex quantum information tasks.

A quantum network refers to a vast configuration of individual quantum nodes communicating with each other via quantum secure channels (Figure 5.3). The fundamental components required for realizing a quantum network are quantum communication channels and quantum interconnects.

Quantum communication channel

A quantum communication channel refers to a physical medium or link through which quantum information is transmitted between quantum systems or nodes. Unlike classical communication channels, which transmit classical bits of information, quantum communication channels enable the transfer of quantum bits or qubits. Quantum communication channels must meet certain requirements to ensure the faithful transmission of quantum information while preserving its delicate quantum properties. These requirements include: minimizing noise and disturbances to preserve the coherence of qubits during transmission and high-fidelity transmission, meaning that the received qubits should be as close as possible to the original qubits sent. Photonics channels, which include fiber optics and free space channels, are one of the best possible platforms to establish quantum links between distant quantum nodes:

- Free space links: In certain scenarios, quantum information can be transmitted through free space using laser beams or photon pairs sent through the atmosphere

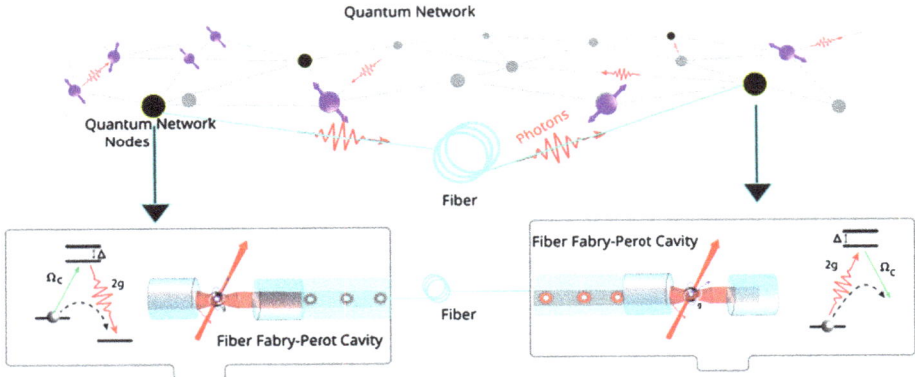

Figure 5.3: A cartoon depiction of a fiber-based quantum network that will interconnect different quantum nodes via fiber channels. In such a scenario, fiber-based light-matter interfaces, e. g., quantum memories, single-photon sources and quantum computing processors, will be crucial for directly integrating quantum interconnects into the network. The single-atom-based fiber-cavity system (inset pictures) has already been shown to fulfill the essential criteria for realizing quantum networks [6, 22]. Typically, a λ-level scheme of an atom is chosen to interface photons in such systems. Photons arrive along the cavity axis from one of the cavity mirrors and couple with strength $2g$ via cavity, and a control laser is used from the side with strength Ω_c for controlled interaction.

or space. Free space quantum communication is particularly useful for satellite-based quantum networks. QKD-based protocols and entanglement distribution have been shown for thousands of kilometers using free space and satellite-based networks [18].

– Fiber optics: Optical fibers are typically used to transmit quantum information over long distances. Single photons or entangled photon pairs can be sent through fiber optic cables, enabling point-to-point quantum communication. However, optical fibers are only suitable for sending optical qubits up to distances of a few hundred kilometers, limited by the typical losses of an optical signal. To enhance long-distance communication using optical fibers as quantum channels, quantum repeater protocols that involve various quantum interconnects are required.

Both fiber and free-space links have their own advantages and disadvantages. For example, optical fiber channels offer better stability and alignment compared to free space channels since the fibers provide a fixed path for the photons, while free space quantum communication channels offer long-distance capabilities. A futuristic quantum network might involve a combination of both approaches.

Fiber-based networks are expected to be used for distances up to a few hundred kilometers. The typical losses of about 0.2 dB/km for the optical signal inside the fiber, limits the qubits transfer to a few hundred kilometers. To further improve long-distance communication using optical fiber channels, requires quantum repeater protocols and quantum interconnects.

Quantum interconnects

Quantum interconnects are specialized components or technologies that enable seamless integration and communication between different quantum systems or nodes. They serve as bridges or interfaces between different types of quantum platforms, allowing for the transfer of quantum information, entanglement, or coherent states. Quantum memories are one of the most crucial quantum interconnects, which are required for long-distance quantum networks. They are also the essential components of quantum repeaters at intermediate quantum nodes.

The essential function of a quantum memory is to perform interconversion of flying photonic qubits into long-lived stationary qubits and on-demand release with high fidelity. As mentioned in Chapter 4, the implementation of quantum memory relies on effectively controlling the interaction between light and matter. This can be achieved by utilizing an optical resonator in conjunction with a suitable matter system, enabling the operation of quantum memory. One advantageous approach is the use of fiber-based resonators, which facilitate the interface between matter qubits and photons and offer natural compatibility with fiber networks. Researchers have demonstrated quantum memory operation in fiber-based resonators for various systems such as single atoms and ions, quantum dots, nitrogen-vacancy centers in diamonds, and other systems. While a future quantum network may exploit specific advantages from different systems and potentially involve a hybrid quantum network, a fiber-based system is a promising approach to achieve compatibility in quantum interconnect networks. Significant research progress has already been made in demonstrating the major functionalities required for general quantum interconnects.

5.3 Fiber-network compatible quantum technology

As we have discussed in the previous section, fiber-based quantum networks are one of the prominent approaches for distributed quantum information processing. By utilizing optical fibers as communication channels, it becomes feasible to seamlessly integrate quantum interconnects if they are fiber-based, for example, matter-based quantum states interface with fiber-based resonators. This aspect has motivated extensive research into the development of fiber cavity-based systems for various tasks:

– Single-photon sources: The development of fiber-coupled single-photon sources with turn-key operation is one of the major research activities. Single-photon sources serve as vital quantum interconnects within the network, and the desired characteristics include high purity, high indistinguishability, high emission rate, and compatibility with fiber-coupled output. To achieve these goals, various techniques are being explored for the design and fabrication of quantum dots, NV centers, and other platforms, in combination with fiber coupling techniques. For instance, approaches such as evanescent coupling to nanofibers with on-chip microcavities, the realization of fiber Fabry–Perot cavities with emitters embedded in one of the reflectors or coupled to the cavity [17], and the development of new coupler techniques directly to single-mode fibers are being pursued.

In addition, atom and ion-based systems have shown promise as excellent sources of single photons. The ability to cool and trap atoms and ions with respect to the mode of an optical fiber allows a very high degree of control over single-photon emission. However, complex techniques of laser cooling and trapping in a vacuum are currently a challenge for scaling these systems.

Continuous progress on fiber-based resonators and coupling techniques will be pivotal for progress in achieving plug-and-play kind of single photon sources. Such sources will greatly enhance the capabilities and functionality of quantum networks.

– Quantum memories: Quantum memories allow coherent and reversible interconversion of photonic quantum states into matter qubit states, strong light-matter interactions with long-lived atomic states is a must requirement. Once again, fiber-based resonators are one of the platforms for strong coupling of various transitions in atoms, ions, quantum dots, NV centers, and other matter qubits as discussed in Chapter 4.

– Quantum networks: Small and medium distance fiber-based quantum networks have been demonstrated and are continuously improved. To just give an idea of the fiber-based quantum channels' capability for transferring quantum information and establishing quantum networks, there have been fiber-based QKD demonstrations using prepare and measure techniques, which do not involve quantum memory operations. For example, the fiber-based QKD for a 421 km fiber length was demonstrated in [5], which is further improved to 830 km using better fibers and better stabilization techniques [27].

Quantum memory-enhanced simple quantum networks [1] in their fundamental working principles, have been shown for entanglement distribution across different physical systems. This includes trapped atoms [13, 6] and ions [19, 24], color centers in diamonds [3, 15, 14, 21], quantum dots [9, 25], and solid-state spin systems in nanophotonic diamond resonators [4].

– Quantum gates: Fiber-based quantum network nodes and the essential fundamental components, which include interfaces capable of sending, receiving, and storing single photons, have been demonstrated in cavity-based systems [6, 22, 4, 26]. The more complex networks involving local quantum information processing require quantum gate operations. Optical cavity coupled to matter systems allows to realize such quantum gates [8, 12, 29, 23]. Therefore, the fiber-based optical cavity system, in principle, allows for the realization of all ingredients for a full-scale quantum network. Further research in improving quantum gate efficiency with scalable schemes with fiber-compatible technology may allow for large-scale distributed quantum processing.

Fiber-based quantum interconnects and quantum networks are expected to play a crucial role in the future of quantum communication and information processing. Advances in fiber technology, such as low-loss fibers and improved signal amplification techniques, will contribute to extending the reach of quantum links, enabling communication over longer distances. Fiber-based quantum networks will be seamlessly integrated with existing classical communication infrastructure. This integration will leverage the extensive fiber optic networks already in place, enabling the coexistence of classical and quantum information transmission. Hybrid classical-quantum networks will provide a bridge between classical and quantum systems, facilitating the transition toward large-scale quantum communication and computation.

Also, further development and optimization for fiber-based light-matter interfaces will enable the distribution of entanglement over much greater distances, potentially enabling global-scale quantum communication networks. Fiber-based quantum networks will continue to evolve toward scalable and robust architectures. Network topologies and routing protocols will be designed to efficiently connect quantum nodes and enable secure and efficient communication among them. This will involve optimizing resource allocation, developing fault-tolerant techniques, and exploring hybrid network designs that combine different types of quantum platforms.

Hopefully, fiber-based optical resonators will continue to drive progress in quantum technology. Advanced techniques of integrated resonators in a chip and photonic-integrated technology seem to be the appropriate technology for scaling and integration of large-scale quantum networks.

Bibliography

[1]   M. Alshowkan, B. P. Williams, P. G. Evans, N. S. V. Rao, E. M. Simmerman, H.-H. Lu, N. B. Lingaraju, A. M. Weiner, C. E. Marvinney, Y.-Y. Pai, B. J. Lawrie, N. A. Peters, and J. M. Lukens Reconfigurable quantum local area network over deployed fiber. *PRX Quantum*, 2:040304, Oct 2021. https://doi.org/ 10.1103/PRXQuantum.2.040304. URL https://link.aps.org/doi/10.1103/PRXQuantum.2.040304.

[2]   C. H. Bennett and G. Brassard. Quantum cryptography: Public key distribution and coin tossing. *Theoretical Computer Science*, 560:7–11, 2014. ISSN 0304-3975. https://doi.org/10.1016/j.tcs.2014.05.025. URL https://www.sciencedirect.com/science/article/pii/S0304397514004241. Theoretical Aspects of Quantum Cryptography – celebrating 30 years of BB84.

[3]   H. Bernien, B. Hensen, W. Pfaff, G. Koolstra, M. S. Blok, L. Robledo, T. H. Taminiau, M. Markham, D. J. Twitchen, L. Childress, and R. Hanson. Heralded entanglement between solid-state qubits separated by three metres. *Nature*, 497(7447):86–90, May 2013. ISSN 1476-4687. https://doi.org/ 10.1038/nature12016.

[4]   M. K. Bhaskar, R. Riedinger, B. Machielse, D. S. Levonian, C. T. Nguyen, E. N. Knall, H. Park, D. Englund, M. Lončar, D. D. Sukachev, and M. D. Lukin. Experimental demonstration of memory-enhanced quantum communication. *Nature*, 580(7801):60–64, Apr 2020. ISSN 1476-4687. https://doi.org/ 10.1038/s41586-020-2103-5.

[5]   A. Boaron, G. Boso, D. Rusca, C. Vulliez, C. Autebert, M. Caloz, M. Perrenoud, G. Gras, F. Bussières, M.-J. Li, D. Nolan, A. Martin, and H. Zbinden. Secure quantum key distribution over 421 km of optical fiber. *Physical Review Letters*, 121:190502, Nov 2018. https://doi.org/10.1103/PhysRevLett.121.190502. URL https://link.aps.org/doi/10.1103/PhysRevLett.121.190502.

[6]   M. Brekenfeld, D. Niemietz, J. D. Christesen, and G. Rempe. A quantum network node with crossed optical fibre cavities. *Nature Physics*, 16(6):647–651, Jun 2020. ISSN 1745-2481. https://doi.org/10.1038/s41567-020-0855-3.

[7]   H.-J. Briegel, W. Dür, J. I. Cirac, and P. Zoller. Quantum repeaters: The role of imperfect local operations in quantum communication. *Physical Review Letters*, 81:5932–5935, Dec 1998. https://doi.org/10.1103/PhysRevLett.81.5932. URL https://link.aps.org/doi/10.1103/PhysRevLett.81.5932.

[8]   S. Daiss, S. Langenfeld, S. Welte, E. Distante, P. Thomas, L. Hartung, O. Morin, and G. Rempe. A quantum-logic gate between distant quantum-network modules. *Science*, 371(6529):614–617, 2021. https://doi.org/10.1126/science.abe3150. URL https://www.science.org/doi/abs/10.1126/science.abe3150.

[9]   A. Delteil, Z. Sun, W.-b. Gao, E. Togan, S. Faelt, and A. Imamoğlu. Generation of heralded entanglement between distant hole spins. *Nature Physics*, 12(3):218–223, Mar 2016. ISSN 1745-2481. https://doi.org/10.1038/nphys3605.

[10] L.-M. Duan, M. D. Lukin, J. I. Cirac, and P. Zoller. Long-distance quantum communication with atomic ensembles and linear optics. *Nature*, 414(6862):413–418, Nov 2001. ISSN 1476-4687. https://doi.org/10.1038/35106500.

[11] N. Gisin, G. Ribordy, W. Tittel, and H. Zbinden. Quantum cryptography. *Reviews of Modern Physics*, 74:145–195, Mar 2002. https://doi.org/10.1103/RevModPhys.74.145. URL https://link.aps.org/doi/10.1103/RevModPhys.74.145.

[12] B. Hacker, S. Welte, G. Rempe, and S. Ritter. A photon–photon quantum gate based on a single atom in an optical resonator. *Nature*, 536(7615):193–196, Aug 2016. ISSN 1476-4687. https://doi.org/10.1038/nature18592.

[13] J. Hofmann, M. Krug, N. Ortegel, L. Gérard, M. Weber, W. Rosenfeld, and H. Weinfurter. Heralded entanglement between widely separated atoms. *Science*, 337(6090):72–75, 2012. https://doi.org/10.1126/science.1221856. URL https://www.science.org/doi/abs/10.1126/science.1221856.

[14] P. C. Humphreys, N. Kalb, J. P. J. Morits, R. N. Schouten, R. F. L. Vermeulen, D. J. Twitchen, M. Markham, and R. Hanson. Deterministic delivery of remote entanglement on a quantum network. *Nature*, 558(7709):268–273, Jun 2018. ISSN 1476-4687. https://doi.org/10.1038/s41586-018-0200-5.

[15] N. Kalb, A. A. Reiserer, P. C. Humphreys, J. J. W. Bakermans, S. J. Kamerling, N. H. Nickerson, S. C. Benjamin, D. J. Twitchen, M. Markham, and R. Hanson. Entanglement distillation between solid-state quantum network nodes. *Science*, 356(6341):928–932, 2017. https://doi.org/10.1126/science.aan0070. URL https://www.science.org/doi/abs/10.1126/science.aan0070.

[16] H. J. Kimble. The quantum internet. *Nature*, 453(7198):1023–1030, Jun 2008. ISSN 1476-4687. https://doi.org/10.1038/nature07127.

[17] A. Kuhn and D. Ljunggren†. Cavity-based single-photon sources. *Contemporary Physics*, 51(4):289–313, 2010. https://doi.org/10.1080/00107511003602990.

[18] C.-Y. Lu, Y. Cao, C.-Z. Peng, and J.-W. Pan. Micius quantum experiments in space. *Reviews of Modern Physics*, 94:035001, Jul 2022. https://doi.org/10.1103/RevModPhys.94.035001. URL https://link.aps.org/doi/10.1103/RevModPhys.94.035001.

[19] D. L. Moehring, P. Maunz, S. Olmschenk, K. C. Younge, D. N. Matsukevich, L.-M. Duan, and C. Monroe. Entanglement of single-atom quantum bits at a distance. *Nature*, 449(7158):68–71, Sep 2007. ISSN 1476-4687. https://doi.org/10.1038/nature06118.

[20] S. Muralidharan, L. Li, J. Kim, N. Lütkenhaus, M. D. Lukin, and L. Jiang. Optimal architectures for long distance quantum communication. *Scientific Reports*, 6(1):20463, Feb 2016. ISSN 2045-2322. https://doi.org/10.1038/srep20463.

[21] M. Pompili, S. L. N. Hermans, S. Baier, H. K. C. Beukers, P. C. Humphreys, R. N. Schouten, R. F. L. Vermeulen, M. J. Tiggelman, L. dos Santos Martins, B. Dirkse, S. Wehner, and R. Hanson. Realization of a multinode quantum network of remote solid-state qubits. *Science*, 372(6539):259–264, 2021. https://doi.org/10.1126/science.abg1919. URL https://www.science.org/doi/abs/10.1126/science.abg1919.

[22] A. Reiserer and G. Rempe. Cavity-based quantum networks with single atoms and optical photons. *Reviews of Modern Physics*, 87:1379–1418, Dec 2015. https://doi.org/10.1103/RevModPhys.87.1379. URL https://link.aps.org/doi/10.1103/RevModPhys.87.1379.

[23] A. Reiserer, N. Kalb, G. Rempe, and S. Ritter. A quantum gate between a flying optical photon and a single trapped atom. *Nature*, 508(7495):237–240, Apr 2014. ISSN 1476-4687. https://doi.org/10.1038/nature13177.

[24] L. J. Stephenson, D. P. Nadlinger, B. C. Nichol, S. An, P. Drmota, T. G. Ballance, K. Thirumalai, J. F. Goodwin, D. M. Lucas, and C. J. Ballance. High-rate, high-fidelity entanglement of qubits across an elementary quantum network. *Physical Review Letters*, 124:110501, Mar 2020. https://doi.org/10.1103/PhysRevLett.124.110501. URL https://link.aps.org/doi/10.1103/PhysRevLett.124.110501.

[25] R. Stockill, M. J. Stanley, L. Huthmacher, E. Clarke, M. Hugues, A. J. Miller, C. Matthiesen, C. Le Gall, and M. Atatüre. Phase-tuned entangled state generation between distant spin qubits. *Physical Review Letters*, 119:010503, Jul 2017. https://doi.org/10.1103/PhysRevLett.119.010503. URL https://link.aps.org/doi/10.1103/PhysRevLett.119.010503.

[26] M. Uphoff, M. Brekenfeld, G. Rempe, and S. Ritter. An integrated quantum repeater at telecom wavelength with single atoms in optical fiber cavities. *Applied Physics B*, 122(3):46, Mar 2016. ISSN 1432-0649. https://doi.org/10.1007/s00340-015-6299-2.

[27] S. Wang, Z.-Q. Yin, D.-Y. He, W. Chen, R.-Q. Wang, P. Ye, Y. Zhou, G.-J. Fan-Yuan, F.-X. Wang, Y.-G. Zhu, P. V. Morozov, A. V. Divochiy, Z. Zhou, G.-C. Guo, and Z.-F. Han. Twin-field quantum key distribution over 830-km fibre. *Nature Photonics*, 16(2):154–161, Feb 2022. ISSN 1749-4893. https://doi.org/10.1038/s41566-021-00928-2.

[28] S. Wehner, D. Elkouss, and R. Hanson. Quantum internet: A vision for the road ahead. *Science*, 362(6412):eaam9288, 2018. https://doi.org/10.1126/science.aam9288. URL https://www.science.org/doi/abs/10.1126/science.aam9288.

[29] S. Welte, B. Hacker, S. Daiss, S. Ritter, and G. Rempe. Photon-mediated quantum gate between two neutral atoms in an optical cavity. *Physical Review X*, 8:011018, Feb 2018. https://doi.org/10.1103/PhysRevX.8.011018. URL https://link.aps.org/doi/10.1103/PhysRevX.8.011018.

[30] F. Xu, X. Ma, Q. Zhang, H.-K. Lo, and J.-W. Pan. Secure quantum key distribution with realistic devices. *Reviews of Modern Physics*, 92:025002, May 2020. https://doi.org/10.1103/RevModPhys.92.025002. URL https://link.aps.org/doi/10.1103/RevModPhys.92.025002.

A Resonator Stabiltiy

In Chapter 2, we have discussed the optical resonators, in particular, Fabry-Perot cavities for confining light. The stability of an optical resonator is also discussed there, which is an important consideration in laser and optical system design. The stability of an optical resonator refers to the conditions under which light can stably oscillate within the resonator without diverging. Therefore, the stability parameter is important in deciding whether the beam can be confined with a chosen cavity geometry and expressed as:

$$g = g_1 g_2$$

where, $g_i = 1 - \frac{L}{R_i}$, $i = 1, 2$ with R_i defined as the radius of curvature of the mirrors and L is the cavity length. There are several commonly used methods for analyzing the stability of optical resonators, with the most well-known being the ABCD matrix method and the stability criteria derived from it. To obtain a stability criterion for a resonator, we will first quickly defined the ABDC matrix used for light propagation through optical elements.

A.1 ABCD Matrix

The ABCD matrix, also known as the "ray transfer matrix", is a mathematical tool used in optics to describe the transformation of optical beams during propagation. The ABCD matrix relates the incoming and outgoing light rays' positions and angles, allowing for the analysis of optical systems involving lenses, mirrors, and other optical components.

The general ABCD matrix for a simple optical system with a single optical element is represented as:

Table A.1: ABCD matrices for some of the optical elements. Here, n_1 and n_2 are refractive indices of the medium on the two sides of the interface.

Optical Element	ABCD Matrix
Free Space (propagation distance d)	$\begin{bmatrix} 1 & d \\ 0 & 1 \end{bmatrix}$
Thin Lens (focal length f)	$\begin{bmatrix} 1 & 0 \\ -\frac{1}{f} & 1 \end{bmatrix}$
Curved Mirror (radius of curvature R)	$\begin{bmatrix} 1 & 0 \\ -\frac{2}{R} & 1 \end{bmatrix}$
Refraction at a flat interface	$\begin{bmatrix} 1 & 0 \\ 0 & \frac{n_1}{n_2} \end{bmatrix}$

https://doi.org/10.1515/9783110636260-006

When light propagates through a series of optical elements, one can find the overall transformation by multiplying the individual ABCD matrices for each element in the sequence.

A.1.1 Derivation Stability conditon

For a two-mirror optical resonator consisting of two mirrors, separated by a distance L, one can write the ABCD matrix for one round trip within the resonator as follows (using Table A.1):

$$M = \begin{pmatrix} (2g_1g_2 - 1) & 2g_2L \\ \frac{2g_1(g_1g_2-1)}{L} & (2g_1g_2 - 1) \end{pmatrix}$$

For a stable resonator, the optical beam should be confined and not escape after many round trips, which is the case when the magnitude of the eigenvalues of the matrix M is less than one. Assming a lossless resonator, i. e., $\det(M) = 1$, the product of the eigenvalues has to be 1. We can write the eigenvalue equation for M as:

$$\det(M - \lambda \cdot I) = \det \begin{vmatrix} (2g_1g_2 - 1) - \lambda & 2g_2L \\ \frac{2g_1(g_1g_2-1)}{L} & (2g_1g_2 - 1) - \lambda \end{vmatrix} = 0$$

The eigen value equation then becomes:

$$\lambda^2 - 2(2g_1g_2 - 1)\lambda + 1 = 0.$$

The eigen values can now be written as:

$$\lambda_{1,2} = (2g_1g_2 - 1) \pm \sqrt{(2g_1g_2 - 1)^2 - 1}.$$

Therefore, the stability criterion for a stable two mirror resonator, which is product of eigenvalues equal to 1, means that the eigenvalues have to be complex conjugate pairs, i. e.:

$$|2g_1g_2 - 1| \le 1.$$

In conclusion, an optical resonator is in a stable configuration to confine light if the geometrical parameters of the resonator satisfy the following condition:

$$0 \le g_1 \cdot g_2 \le 1$$

B Cylindrical symmetry and Bessel function

Cylindrical symmetry refers to a situation where a physical system or a problem possesses rotational symmetry around an axis. In other words, the properties of the system are independent of the angle of rotation around the central axis. One example of this is the case of light propagation in an optical fiber (see Chapter 1). The light propagation axis is the symmetry axis in this case.

B.1 Bessel function

Bessel functions are a family of solutions to Bessel's differential equation, which appears in problems with radial symmetry, such as those encountered in cylindrical coordinates. The Bessel differential equation is given by

$$x^2 \frac{d^2y}{dx^2} + x \frac{dy}{dx} + (x^2 - a^2)y = 0, \tag{B.1}$$

where x is the independent variable (often related to the radial coordinate in cylindrical systems) and a is a constant. One of the methods for obtaining a solution to the Bessel differential equation is by assuming a solution in terms of a power series solution:

$$y = A_0 + A_1x + A_2x^2 + A_3x^3 + \cdots \tag{B.2}$$

Substituting above form of y into equation (B.1) and then equating the coefficient of the same power of x leads to the functional form of the solutions. One can build the solutions starting from the lowest order, i. e., $n = 0, 1, 2, 3 \ldots$ and power series form of the Bessel functions ($J_n(x)$ and $Y_n(x)$) are obtained.

Bessel functions are defined in different ways depending on the specific problem. For light propagation in a cylindrical waveguide, we encounter Bessel functions of the first and second kind, denoted by $J_n(x)$ (for integer $a = n$) or $Y_n(x)$ (for noninteger $a = n$); and modified Bessel functions of the first and second kind, denoted by $I_a(x)$ and $J_a(x)$. The lower subscript in the function represents the order of the Bessel function. These functions are used to describe various physical phenomena, such as heat conduction in a cylindrical object, the behavior of electromagnetic waves in a cylindrical waveguide, and the radial modes of vibration in circular membranes.

Bessel functions of the first kind, $J_n(x)$, are well-behaved at the origin ($x = 0$) and are generally used when considering physical problems bounded at the origin. Bessel functions of the second kind, $Y_n(x)$, diverge at the origin and are more appropriate for unbounded problems or problems where the origin is not a concern. Series expansion of $J_n(x)$ is given by

https://doi.org/10.1515/9783110636260-007

$$J_n(x) = \sum_{m=0}^{\infty} \frac{(-1)^m}{m!\,\Gamma(m+n+1)} \left(\frac{x}{2}\right)^{2m+n}, \tag{B.3}$$

where $\Gamma(x)$ is the gamma function. The Bessel function of the second kind that has a singularity at the origin can be represented as follows:

$$Y_n(X) = \frac{J_n(x)\cos(n\pi) - J_{-n}(x)}{\sin(n\pi)} \tag{B.4}$$

The Bessel function plots look similar to the oscillatory sine and cosine functions with decay functions. These functions are crucial in solving partial differential equations with cylindrical symmetry, as they often appear as solutions due to the radial nature of the coordinate system.

B.1.1 Modified Bessel function

Bessel functions for complex arguments are called modified Bessel functions. Modified Bessel functions of the first and second kinds are represented by $I_a(x)$ and $K_a(x)$. They can be expressed in terms of $J_a(x)$:

$$I_a(x) = i^{-a} J_a(ix) \tag{B.5}$$

$$K_a(x) = \frac{\pi}{2} \frac{I_{-a}(x) - I_a(x)}{\sin a\pi} \tag{B.6}$$

Modified Bessel functions are exponentially growing or decaying functions.

B.1.2 Helmholtz equation in cylindrical coordinates

$$\begin{aligned}
\Delta^2 \varphi &= \frac{1}{\rho}\frac{\partial}{\partial \rho}\left(\rho \frac{\partial \varphi}{\partial \rho}\right) + \frac{1}{\rho^2}\frac{\partial^2 \varphi}{\partial \phi^2} + \frac{\partial^2 \varphi}{\partial z^2} \\
&= \frac{\partial^2 \varphi}{\partial \rho^2} + \frac{1}{\rho}\frac{\partial \varphi}{\partial \rho} + \frac{1}{\rho^2}\frac{\partial^2 \varphi}{\partial \phi^2} + \frac{\partial^2 \varphi}{\partial z^2}
\end{aligned} \tag{B.7}$$

The Helmholtz wave equation $\Delta^2 \varphi = 0$ can be written as

$$\frac{\partial^2 \varphi}{\partial \rho^2} + \frac{1}{\rho}\frac{\partial \varphi}{\partial \rho} + \frac{1}{\rho^2}\frac{\partial^2 \varphi}{\partial \phi^2} + \frac{\partial^2 \varphi}{\partial z^2} = 0 \tag{B.8}$$

The above equation can be solved by the method of variable separation, considering

$$\varphi(\rho, \phi, z) = R(\rho)\Phi(\phi)Z(z). \tag{B.9}$$

The Helmholtz equation separates into three independent equations as follows:

$$\frac{d^2Z(z)}{dz^2} - k^2 Z = 0, \tag{B.10}$$

$$\frac{d^2\Phi(\phi)}{d\phi^2} + v^2\Phi = 0, \tag{B.11}$$

$$\frac{d^2R(\rho)}{d\rho^2} + \frac{1}{\rho}\frac{dR(\rho)}{d\rho} + \left(k^2 - \frac{v^2}{\rho^2}\right)R(\rho) = 0. \tag{B.12}$$

The first two equations are standard second-order differential equations, while the last equation is the Bessel differential equation.

B.2 Wave equation in cylindrical coordinates

In cylindrical coordinates (r, θ, z), the wave equation is a partial differential equation that describes wave propagation in systems with cylindrical symmetry. The wave equation in cylindrical coordinates is given by

$$\nabla^2\Psi - (1/c^2)\frac{\delta^2\Psi}{\delta t^2} = 0 \tag{B.13}$$

where $\Psi = \Psi(r, \theta, z, t)$ is the scalar wave function, representing, for example, the displacement or pressure amplitude of the wave at position (r, θ, z) and time t. c is the wave propagation speed in the medium. ∇^2 is the Laplacian operator in cylindrical coordinates given by:

$$\nabla^2 = \frac{1}{r}\frac{\delta}{\delta r}\left(r\frac{\delta}{\delta r}\right) + \left(\frac{1}{r^2}\right)\frac{\delta^2}{\delta\theta^2} + \frac{\delta^2}{\delta z^2}. \tag{B.14}$$

To solve the wave equation in cylindrical coordinates, we can use the method of separation of variables. We assume that the wave function $\psi(r, \theta, z, t)$ can be expressed as a product of three functions, each depending on only one coordinate:

$$\Psi(r, \theta, z, t) = R(r)\,\Theta(\theta)\,Z(z)\,T(t). \tag{B.15}$$

Substituting this into the wave equation and rearranging, we get

$$\frac{1}{c^2}\frac{d^2T}{dt^2} = \frac{1}{R}\frac{d}{dr}\left(r\frac{dR}{dr}\right) + \frac{1}{r^2}\frac{d^2\Theta}{d\theta^2} + \frac{1}{Z}\frac{d^2Z}{dz^2} \tag{B.16}$$

Now, since each side of the equation depends on different independent variables, both sides must be equal to a constant, which we will denote as $-k^2$:

$$\frac{1}{c^2}\frac{d^2T}{dt^2} = -k^2; \quad \left(\frac{1}{R}\frac{d}{dr}\left(r\frac{dR}{dr}\right) + \frac{1}{r^2}\frac{d^2\Theta}{d\theta^2} + \frac{1}{Z}\frac{d^2Z}{dz^2}\right) = -k^2 \tag{B.17}$$

The last term of the equation depends only on z, and therefore, from partial derivative of the equation with respect to z indicates this term to be a constant. We can consider a constant a such that:

$$\frac{d^2Z}{dz^2} + a^2Z = 0. \tag{B.18}$$

Therefore, the longitudinal component equation (along the z-axis) gives the general solution,

$$Z(z) = A\cosh(az) + B\sinh(az), \tag{B.19}$$

where A and B are constants determined by the boundary conditions.

Once we reorganize the terms in equation (B.17), we get three separate ordinary differential equations (ODEs) to solve, each with its own constants:

Time component equation:

$$\frac{d^2T}{dt^2} + c^2k^2T = 0 \tag{B.20}$$

Radial component equation:

$$r^2\frac{d^2R}{dr^2} + r\frac{dR}{dr} - k_1^{\,2}R = 0 \tag{B.21}$$

Azimuthal component equation:

$$\frac{d^2\Theta}{d\theta^2} + m^2\Theta = 0. \tag{B.22}$$

The solutions to the azimuthal component equation, equation (B.22), are the familiar periodic functions of the azimuthal angle θ, which are the solutions to the angular part of the wave equation. These solutions are given by

$$\Theta(\theta) = Ce^{(im\theta)} \tag{B.23}$$

where m is an integer, known as the azimuthal mode number, and C is a complex number.

The radial component equation, equation (B.21), is a Bessel equation, and its solutions are Bessel functions of the first kind ($J_m(k_1r)$) and Bessel functions of the second kind ($Y_m(k_1r)$). The general solution for the radial part of the wave equation is a linear combination of these two types of Bessel functions:

$$R(r) = DJ_m(k_1r) + EY_m(k_1r) \tag{B.24}$$

where D and E are constants determined by boundary conditions.

The time component equation is a simple harmonic oscillator equation, and its general solution is a linear combination of sines and cosines:

$$T(t) = F\cos(ckt) + G\sin(ckt) \tag{B.25}$$

where F and G are constants determined by initial conditions.

To get the complete solution for the wave equation in cylindrical coordinates, we combine the solutions for each component:

$$\Psi(r, \theta, z, t) = (A\cosh(az) + B\sinh(az))(Ce^{(im\theta)})(DJ_m(k_1 r) + EY_m(k_1 r))$$
$$\times (F\cos(ckt) + G\sin(ckt)) \tag{B.26}$$

This is the general solution to the wave equation in cylindrical coordinates, which describes wave propagation with cylindrical symmetry and can be applied to various physical systems, such as sound waves in a cylindrical pipe or electromagnetic waves in cylindrical waveguides. The specific values of the constants $A, B, C, D, E, F, G, m, k, a$ depend on the boundary and initial conditions of the problem being solved.

C Atom-Light Interaction

A comprehensive quantum mechanical description of atom-light interaction is achieved through the field of cavity quantum electrodynamics (CQED). In this field, atoms and light are both treated as quantized entities, reflecting their inherent quantum nature. Therefore, CQED requires the quantization of electromagentic field also.

C.1 Quantization of Electromagnetic field

The quantum mechanical picture of an interaction between a photon and an atom requires the quantization of the light field. Assuming an electromagnetic field confined in a three-dimensional volume, one can follow the following steps for the quantization of light. First, we can consider a finite box of dimensions $L \times L \times L$, which can be eventually extended to the case where $L \to \infty$, and try to find the allowed solution within the bounded region. The Maxwell equations for this confined space, where the charges and currents are absent, can be written as follows:

$$\nabla \cdot \mathbf{E} = 0 \tag{C.1}$$

$$\nabla \times \mathbf{E} = -\frac{\partial \mathbf{B}}{\partial t} \tag{1.1}$$

$$\nabla \cdot \mathbf{B} = 0 \tag{C.2}$$

$$\nabla \times \mathbf{B} = \frac{1}{c}\frac{\partial \mathbf{E}}{\partial t} \tag{1.2}$$

where the speed of light emerges as $c = 1/\sqrt{\epsilon_0 \mu_0}$. It is common to relate the both the electric and magnetic field amplitudes to a vector potential:

$$\mathbf{E} = -\frac{\partial \mathbf{A}}{\partial t} \tag{C.3}$$

$$\mathbf{B} = \nabla \times \mathbf{A} \tag{C.4}$$

under a condition known as Coulomb gauge:

$$\nabla \cdot \mathbf{A} = 0.$$

Using these definitions in terms of vector potential $\mathbf{A}$ results in the following wave equation:

$$\nabla^2 \mathbf{A}(\mathbf{r}, t) = \frac{1}{c^2}\frac{\partial^2 \mathbf{A}(\mathbf{r}, t)}{\partial t^2}$$

with solutions that are split into positive and negative components.

https://doi.org/10.1515/9783110636260-008

$$\mathbf{A}^+(\mathbf{r}, t) = i \sum_{\mathbf{k}} c_{\mathbf{k}} u_{\mathbf{k}}(\mathbf{r}) e^{-i\omega_k t}$$

and $\mathbf{A}^{(+)}(\mathbf{r}, t) = [\mathbf{A}^{(-)}(\mathbf{r}, t)]^*$. We have used identity $\nabla^2 \mathbf{A} = \nabla(\nabla \cdot \mathbf{A}) - \nabla \times (\nabla \times \mathbf{A})$. The coefficients $c_{\mathbf{k}}$ with index $\mathbf{k}$ specify the wave-vector as well as polarization mode. We consider only the linear polarization and ignore the polarization degree. Also, in free space the dispersion relation is simply $\omega_k = ck$, relating the frequency of the mode only to the absolute value of the wave-vector. The wave equation is now simplified to:

$$\left(\nabla^2 + \frac{\omega^2 k^2}{c^2}\right) u_{\mathbf{k}}(\mathbf{r}) = 0$$

Considering the Coulomb gauge condition such that the divergence of individual spatial mode functions is vanishing:

$$\nabla \cdot \mathbf{u}_k(\mathbf{r}) = 0$$

Since these are eigenvectors of the differential operator above, they form a complete orthonormal set:

$$\int_V dr\, \mathbf{u}_{\mathbf{k}}^*(\mathbf{r}) \mathbf{u}_{\mathbf{k}'}(\mathbf{r}) = \delta_{\mathbf{k}\mathbf{k}'}$$

Now, with the periodic boundary conditions, the solutions for the above equation are well-known traveling waves:

$$\mathbf{u}_{\mathbf{k}}(\mathbf{r}) = \frac{1}{L^{3/2}} e^{i\mathbf{k}\cdot\mathbf{r}} \boldsymbol{\epsilon}^{\mathbf{k}}$$

We could have also chosen different boundary conditions, which would have given standing waves for the mode functions. The allowed wave-vectors have components $(n_x, n_y, n_z)2\pi/L$ where the indexes are positive integers. Now, we can finally write the vector potential in terms of a constant defined by the mode volume and mode coefficients $c_{\mathbf{k}}$:

$$\mathbf{A}(\mathbf{r}, t) = i \sum_{\mathbf{k}} \sqrt{\frac{\hbar}{2\epsilon_0 \omega_k V}} \left[c_{\mathbf{k}} e^{i\mathbf{k}\cdot\mathbf{r}} e^{-i\omega_k t} - c_{\mathbf{k}}^* e^{-i\mathbf{k}\cdot\mathbf{r}} e^{i\omega_k t} \right] \boldsymbol{\epsilon}^{\mathbf{k}}.$$

One can now get the full expression for the electric field as:

$$\mathbf{E}(\mathbf{r}, t) = \sum_{\mathbf{k}} \sqrt{\frac{\hbar\omega_k}{2\epsilon_0 V}} \left[c_{\mathbf{k}} e^{i\mathbf{k}\cdot\mathbf{r}} e^{-i\omega_k t} - c_{\mathbf{k}}^* e^{-i\mathbf{k}\cdot\mathbf{r}} e^{i\omega_k t} \right] \boldsymbol{\epsilon}^{\mathbf{k}}.$$

The quantization is now obtained by replacing the c-number amplitudes with operators satisfying the following relations: $[a_{\mathbf{k}}\, a^+{}_{\mathbf{k}'}] = \delta_{\mathbf{k}\mathbf{k}'}$. The expression for the quantized electric field as a sum of negative and positive frequency components is:

Here, the summation over index k represents the sum over different modes of the electromagnetic field. The coupling coefficient is defined as follows:

$$g_{\mathbf{k}} = \sqrt{\frac{\hbar \omega_{\mathbf{k}}}{2\epsilon_0 V}} (\mathbf{d} \cdot \hat{e}_{\mathbf{k}}) \hbar = \frac{E_{\mathbf{k}}(\mathbf{d} \cdot \hat{e}_{\mathbf{k}})}{\hbar}$$

Index

https://doi.org/10.1515/9783110636260-009

www.ingramcontent.com/pod-product-compliance
Lightning Source LLC
Chambersburg PA
CBHW081532220326
41598CB00036B/6402